The History of Science and the New Humanism

The History of Science and the New Humanism

George Sarton

With recollections and reflections by Robert K. Merton

Transaction Books
New Brunswick (U.S.A.) and Oxford (U.K.)

Library of Congress Catalog No: 87-10903
ISBN: 0-88738-703-9
Printed in the United States of America

Library of Congress cataloging-in-Publication Data

Sarton, George, 1884-1956.
The History of science and the new humanism.

Reprint. Originally published: Cambridge, Mass.: Harvard University Press, 1962.
Includes bibliographical references and index.
1. Science--History. 2. Civilization--History. 3. Humanism--20th century. I. Title.
Q125.S25 1987 509 87-10903
ISBN 0-88738-703-9

Contents

N.B. *Lectures I, II, III are the Colver Lectures; Lecture IV is the Elihu Root Lecture.*

TO MY DEAREST FRIEND

E. M. S.

THE MOTHER OF THOSE STRANGE TWINS

MAY AND ISIS

RECOLLECTIONS & REFLECTIONS

George Sarton: Episodic Recollections
by an Unruly Apprentice

By Robert K. Merton

Half a century has raced and stumbled by since I first found myself, as a third-year graduate student in sociology at Harvard, daring to knock on the door of George Sarton's famed workshop-cum-study, Widener 185-189. The reason for taking this daunting step was clear: having elected to try my hand at a dissertation centered on sociologically interesting aspects of the efflorescence of science in seventeenth-century England-a kind of subject not exactly central to sociology back then-it did not seem unreasonable to seek guidance from the acknowledged world dean among historians of science.

Although Emerson Hall, which housed the Department of Sociology, was only a hundred paces from Widener, this was not a short journey. Traffic to the Sarton workshop by denizens, mature or immature, of the newfangled Department of Sociology faced formidable barriers. For one thing, the few graduate students who then had any knowledge of Sarton's scholarly existence took him to be a remote, austere, and awesome presence, so thoroughly dedicated to his scholarship as to be quite unapproachable by the likes of us. Thus do plausible but ill-founded beliefs develop into social realities through the mechanism of the self-fulfilling prophecy. Since this forbidding scholar was bound to be unapproachable, there was plainly small point in trying to approach him. And his subsequently having little to do with graduate students only went to show how inaccessible he actually was.

Buttressing this imputed barrier of personal inaccessibility were the authentic university barriers of departmental organization. The understaffed Department of Sociology, established just three years before, had enlarged its graduate program by reaching out to list research-and-reading courses in a great variety of departments: psychology and economics; government, religion, and philosophy; anthropology and social ethics amongst them. But nary a graduate course in the history of science. This for the best of reasons–Harvard had no autonomous department devoted to that undisciplined subject, nor, for that matter, had any other university. Still, in the preceding academic year, 1932-1933, I had managed to audit the sole lecture course in the field, entitled "History of Science 1. History of the Physical and Biological Sciences." (As you may have begun to suspect, the title HS 1 was an unredeemed promissory note; there was no HS 2 back then.)

The first semester of the course was given by the biochemist and polymath of great note, L.J. Henderson, later described by James Conant as "the first roving professor in Harvard." And rove he did. Not only had he instituted the course in the history of science two decades before, but, in that same year of 1932, he had also instituted his unique graduate "Seminary in Sociology" entitled "Pareto and Methods of Scientific Investigation." The plural "methods" rather than the more familiar and misleading singular, "the scientific methods," also reflected a theme in the first semester of the history of science course as the "pink-whiskered" Henderson engaged in his typically forceful, magisterial exegesis of texts by Hippocrates, Galileo, and Harvey-thus allowing him to expound his conception of the varieties of scientific inquiry. But as Conant confirms, Henderson, like Sarton, would have hooted at the then not uncommon notion that a grounding in the history of science served to sharpen one's capabilities as a scientific investigator.

The second semester of this lone course in the history of science was given by the lecturer, *Dr.* Sarton-decidedly not yet *Professor* Sarton; that title was only to come seven years later, when Sarton was fifty-six, and Conant, as Harvard's president, finally intervened to bring it about. Sarton differed greatly from Henderson in both the style and substance of his teaching. Warmly enthusiastic rather than coldly analytical-in a fashion that plainly irritated Henderson from time to time[1]-Sarton traced expanses of scientific development chiefly through the lives and accomplishments of what he took as prototypal figures in that development. (I gather from I. Bernard Cohen's recent account of that course as it was a few years later that all this remained much the same.[2] Looking back, one is inclined to say that if Henderson still dressed in Edwardian style, Sarton still thought in Edwardian style. Both were thoroughly engaging in their fashion; neither is now readily reproducible.

As a mere graduate student, I knew nothing, of course, about the grim vicissitudes Sarton was experiencing in the determined effort to supplement his own scholarship with institutional arrangements designed to advance the cause of the historiography of science. But here is Conant's retrospection on Sarton's incessant efforts at this time (when Conant was president of the university and a self-declared amateur in the history of science):

> This is not the time or place to summarize the history of Professor Sarton's long years at Harvard, his prodigious scholarship, his editorship of *Isis* and *Osiris*, and his vain attempt during the depression years to persuade Harvard or any other university to endow what he considered a minimal department of the history of science. That we are meeting here tonight with a teaching staff in the history of science at Harvard in active service, that a flourishing undergraduate and graduate field of study in history and science has long been characteristic of this University are some of the fruits of George Sarton's long uphill struggle to make the history of science an important part of the American scene.[3]

But this public statement does not fully reflect Conant's complex image of Sarton back in the 1930s, which evidently was, and long remained, ambivalent. That ambivalence was expressed in a letter written almost forty years later regarding the first biographical piece. Arnold Thackray and I published about Sarton: "You are quite right in giving Henderson a key place in your story. I talked to him more than once about Sarton and he reported on his difficulties with this stubborn genius. Henderson often served as an intermediary. He understood how exorbitant were Sarton's demands. Your footnotes 29 and 30 are quite correct. My viewpoint was greatly influenced by Henderson."[4]

But enough about those hard times for George Sarton. In an obviously Tristram Shandy mode, where it takes more time to record life than to live it, I have left my youthful self in the fall of 1933 knocking on the door of that austere scholar's study in Widener, quite determined yet rather fearful of this first face-to-face audience with his august presence. (I say "august presence," for so it seemed to me at the time, although he was then still in his forties, just as I say "first audience" since I had not before had a private session with him, having attended his course only when I could escape from duties as a teaching-and-research assistant to the sociologist Pitirim Sorokin.) On that initial well-remembered occasion, the reputedly unapproachable scholar did not merely invite me into his "tiny book-lined study"; he positively *ushered* me in. Thus began my short, incompleat, and sometimes unruled apprenticeship, followed by an intermittent epistolary friendship that continued until his death in 1956. I began that first audition by telling of my plans for a dissertation already begun. I cannot say that he greeted those plans with conspicuous enthusiasm; instead he mildly suggested that so large a canvas as seventeenth-century English science might be a bit excessive for a novice. But he did not veto the idea. I should describe his response as, at best, ambivalent. Having registered his doubts, he then proceeded to tailor a research course to the needs of the first graduate student to have come to him from the social sciences since his arrival at Harvard some seventeen years before.

I now suspect that the unheralded appearance of a young sociologist-in-the making may have reactivated his own youthful ecumenical vision of transcending disciplinary boundaries. Recall only his vision, full of innocence and hope, of the about-to-be-launched *Isis* as "at once the philosophical journal of the scientists and the scientific journal of the philosophers, the historical journal of the scientists and the scientific journal of the historians, the sociological journal of the scientists and the scientific journal of the sociologists."[5] As one notes, that daunting aspiration called not alone for a philosophy, history, and sociology of science but also for the sciences of philosophy, history, and sociology, all to find suitable expression in this variously ecumenical journal. That aspiration, it might be observed, was not much diminished by the circumstance that two years after its founding in 1912, *Isis* had acquired a world total of 125 subscribers. Of all that I had not the remotest idea when I venturously crossed the threshold of Widener 185, where worked the founder-editor of *Isis* and the author of the newly published monumental two volumes of an *Introduction to the History of Science*, which had managed to make its way from Homer through the thirteenth century in some 2,000 closely printed pages. Since, not quite incidentally, he was also a Harvard lecturer, I was there to ask that this composite personage break through all bureaucratic barriers to establish a research course for a neophyte sociologist.

Happily, Harvard was not in the hands of bureaucratic virtuosos and manifestly that special course was soon arranged; else I would not be thinking back on the devices this early master of the art and craft of the history of science invented to bring that maverick sociologist across academic boundaries into the then hardly institutionalized discipline of the history of science.

And now I undermine credibility by reporting that, during those many years-first as student and apprentice, then as journeyman and junior colleague, and finally as a properly certified scholar in my own right-I do not recall having been seriously irritated by this deeply committed, often impatient, and sometimes difficult scholar. Considering that he has been declared variously exasperating and downright abrasive by early colleagues and later students-I again need instance only his ambivalent advocates L.J. Henderson and James Conant and his student I.B. Cohen-it appears either that I simply lacked the same sensibility or the same range of close, continued interaction, or perhaps, that I have managed to repress, beyond all hope of retrieval except through the deployment of drastic psychoanalytic techniques, a deep underlying irritation that would evoke an intolerable conscious sense of guilt were it allowed to surface. I reject that last plausible hypothesis (be it noted without a betraying excess of protest). It simply doesn't wash.

There is yet another evident hypothesis: that in truth, George Sarton happened to treat me with friendly care, even with solicitude. This is somewhat more plausible. It has the further merit of being in accord not merely with possibly undependable memory traces but with personal documents. From them, the plain fact emerges that I liked and appreciated Sarton even when he was having at me for departures from the Comtean faith, or quite rightly, was reminding me of defections from norms governing the several roles of the scholar, such as my not getting reviews of books or referee reports in on time. Nor is it surprising that I should have remained attached to him, early and late in our evolving relationship. For as I have discovered only now in reliving the history of that relationship for this centenary moment, he had bound me to him–not with any such intent, I believe–by a flow of gifts, freely bestowed, which in their cumulative outcome may have affected my life and work in ways that have little or nothing to do with substantive doctrine or method of inquiry but much to do with discovering the pleasures and joys, as well as the nuisances and pains, of life as a scholar. I now see that he provided an accumulation of advantage,[6] thus leading me to incur a debt that called for a life of continuing work long after the insidious temptations of an easy retirement have been painlessly resisted.

Only now, decades after the events, have I come to recognize the patterned flow of the gifts, material and symbolic, which this ostensibly peripheral mentor bestowed upon me. And should I be exaggerating their import and consequences, as I may be doing in the first flush of their composite discovery, they remain nevertheless as I describe them. But if that large claim of the Sartonian largess is to persuade me, let alone you, they must not rest on vagrant memories-that is, memories without visible means of documentary support. For that reason, I shall draw upon fragments of the correspondence between us, as a basis for the rest of this episodic glimpse into George Sarton's mentorial style.

THE GIFTS

The first gift was his accepting a graduate student drawn from a department of learning in which he took no part. By intimation rather than in so many words, this was on condition that I did not threaten his "disciplined routine" of scholarship or require him to abate "the fury with which he set himself to work."[7] Having made that evident, he went on to provide me with a place in the large workshop adjacent to his small study, which I shared, to a degree, with his secretary, Frances Siegel, and his research associates, the formidable Dr. Alexander Pogo in the field of astronomy and the accommodating Dr. Mary Catherine Welborn in medieval studies. That microenvironment itself constituted a second-order gift, for I learned many now-indeterminate things from that variegated pair of talented associates, albeit through a kind of cognitive osmosis rather than through formal training.

From the beginning, George Sarton did much to help set me on the path of scholarship. He proceeded methodically-he was methodical in most things-to transform me from a graduate student, struggling with preliminary work on a dissertation, into a tyro scholar addressing an international quasi community of scholars in print. This he did first by opening the pages of *Isis* to me. During the next few years, he accepted several articles of mine along with some two dozen signed reviews and another twenty or so entries for its annotated critical bibliographies.

In retrospect, I am persuaded that this initial run of scholarly experience served as both catalyst and exemplare. This I infer from finding that my first batch of published reviews, all eight of them, appeared in *Isis*, and that I soon went on to write a good many articles and reviews for other journals during that period. Moreover, had it not been for a publication schedule noticeably slowed by having *Isis* printed abroad-this by the St. Catherine Press in Bruges-the paper entitled "The Course of Arabian Intellectual Development, 700-1300 A.D.," which was written for *Isis* (in collaboration with Sorokin) might also have been my first article to appear in print. At any rate, I have a note from Sarton, dated just two months after he had admitted me to Widener 185 and addressed to me at the infirmary where I had completed the manuscript. In it, he writes: "I will try to come to see you before I leave" (this, for a needed restful cruise in the West Indies) and then appends the seemingly casual postscript: "Will be delighted to publish your paper in *Isis*."[8] Eighteen months later, it was in print.

The flow of gifts continued. Once equipped with a desk in the Sarton workshop, I was allowed to move freely through the fabulous bibliographic files, asked to serve as referee for the vanishingly few manuscripts with a sociological tinge, and enabled to read selectively in the galleys of the forthcoming contents of *Isis*–the latter a privileged access that would dramatically affect the oral examination on my dissertation. But months before the fateful occasion–indeed, before I had actually completed the dissertation–Sarton is writing me an altogether astonishing letter which reads in its entirety thus:

3510.08[9]

Dear Merton,

I have examined your thesis with great interest and have read much of it. I think it is an excellent piece of work and warmly congratulate you.

The lack of a table of contents–including the chapters written & unwritten–makes it difficult to appreciate the symmetry of the whole structure.

From p. 267 on should in my opinion form a new chapter, Chapter X.

[In the event, it did. And then comes the paragraph with its climactic gift.]

The sincerity of my praise of your work will be best established by my readiness to publish it in *Osiris*, vol. 2 or 3, if H.U. or another agency is ready to share the financial burden and risk with me. This would be the cheapest mode of publication.

[In the event, neither Harvard nor any other agency shared the burden and the risk; my mentor himself provided the functional equivalent of a publishing grant from the nonexistent National Science Foundation. And then Sarton concludes the letter with a manifestly ambivalent judgment.]

The work should be somewhat condensed, notably the religious part–though this might be difficult, as I found no trace of prolixity. [In the event, this part was not condensed in the published version.]

With kind regards & best wishes,
George Sarton

[And then, an afterthought expressing the lifelong Sartonian concern with indexes, about which I shall have more to say.]

In the case of publication an index should be added, but it might be compiled on the page proofs.

A few words about the magnitude of that gift. During my impromptu rather than regularly scheduled sessions with Sarton, he had made it clear that he preferred not to discuss my developing dissertation nor to see the manuscript until it was well-nigh complete. Instead, those sessions were largely given over to his telling an interested listener about the work life of a scholar with a defined mission: about the long frustrated yet continuing aspirations for an institute or a department of the history of science, about the problems of keeping *Isis* intellectually and financially solvent, about the slowly evolving work on the third huge volume of the *Introduction*, about the extraordinary array of requirements for the proper education of an encyclopaedic historian of science (an exceedingly demanding array which was much moderated in his later public statements on the subject), and so on. Thus that letter with its emphatic vote of confidence in my manuscript came without the least prior intimation that he had accepted in the event what he had understandably doubted in the intent as an excessively ambitious subject for a dissertation.

Nor, of course, was that gift wholly symbolic; it had a decidedly practical aspect. Like other newly minted Ph.D.s in those Depression years of the 1930s, I had pretty much assumed that the dissertation would not be published since it would see print only if I should subsidize publication (as I manifestly could not). Then came the Sarton offer, with its contingent-subsidy clause soon removed. I did not refuse that gift, either.

Even so, all this was only prologue to the dissertation defense a short while later. As sponsor and chairman, Pitirim Sorokin was hard put to piece together an appropriate examining committee for this out-of-phase dissertation. The mandatory three members of the understaffed Department of Sociology were Sorokin himself, whom I was assisting in writing the chapters dealing with sociological aspects of scientific discovery and technological invention in his four volume *Social and Cultural Dynamics*; the young instructor Talcott Parsons, still two years away from his masterwork, *The Structure of Social Action*, and with no public identity as a sociologist since he had published only two articles all told, which derived from *his* dissertation; and Carle C. Zimmerman, the rural sociologist Sorokin had brought with him from the University of Minnesota. The fourth member of the committee was George Sarton.

In an action composed of equal measures of deference and prudence, Sorokin invited Sarton to begin the examination. His first question struck me dumb: "Mr. Merton, will you tell us, please, who discovered the greater circulation of the blood?" In a matter of milliseconds, as I now reconstruct it, these anxious thoughts raced through my mind: "What *is* he up to? How can he possibly ask this elementary question? After all, he knows that Henderson[10] has put us through our paces on Harvey and he knows that I've been visiting seventeenth-century England for several years, and he *has* said kind things about the dissertation. How *could* he ask that question? What *is* he up to?"

Rendered utterly desperate by the thought that my mentor was sadistically subjecting me to some arcane test of competence, I launched on this approximate reply (thought not, I suspect, with as orderly a syntax): "Of course, the greater circulation of the blood was discovered by William Harvey. Some claim that it was intimated in his lecture notes of 1616, although he didn't get around to publishing it until 1628, and some maintain that even his *De motu cordis* . . ."-and so on and so on, in the familiar textbook style.

Then came this desperate plunge into irrelevance: "But for historians of science, the recent exitement lies in the new confirmation, called for some years ago in your *Introduction*, that the thirteenth-century Arab physician Ibn al-Nafis did indeed discover the lesser pulmonary circulation, long before its independent discovery first by Servetus and then by Columbo. It should be said, however, that he arrived at the lesser circulation, not through dissection, which was of course taboo in his culture, but on strictly theoretical grounds. Furthermore, . . ."

At this point, George Sarton literally rose to the occasion. With expressive disbelief and enthusiasm, he leapt to his feet, pounded the large library table round which sat the inquisitors and their innocent victim, and exclaimed: "How *could* you know of that confirming evidence? My old friend Max Meyerhof found several manuscripts of the Arabic text in Cairo and published them in a specialized German journal in the history of medicine which you would surely have no reason to read. Later, he sent me a condensed translation for *Isis* where it appears in a belatedly distributed issue. Tell me, how *do* you know of this new evidence?"

For a moment, I was anxious rather than triumphant. Would my examiners from Sociology think that this was all a put-up job between Sarton and myself? Nevertheless, I went on to explain: "But as you know, Dr. Sarton, when I'm at my desk in your Widener workshop, I make a point of reading certain articles in galleys and Meyerhof's happened to be one I read."

The rest of that *rite de passage* known as a doctoral examination was smooth and pleasurable sailing.[11] But to this day, I do not truly know what Sarton had in mind. Having been an examiner on scores of such occasions, I suspect that he introduced that elementary question simply to put me at ease. This imputed intent becomes the more plausible in light of the journal entry on his own doctoral examination. There he writes: "I passed my examination pitifully: the first ordeal toward the doctorate in physical and mathematical sciences. It made a very painful impression on my professors and will do me a lot of harm at the final examination."[12]

Whatever his intent, George Sarton had bestowed another, possibly inadvertent gift. For had he not mystified me by that opening question, I would not have had the occasion or the temerity to tell of Ibn al-Nafis and thus to impress my professors by that display of new-found, distinctly limited, and altogether irrelevant erudition.[13]

Sarton's gifts of publishing the dissertation and getting me off to a grand though unearned start in the examination belong to the class of what the anthropological poet-ethicist Lewis Hyde describes as "threshold gifts." These, he notes, "mark the time of, or act as the actual agents of, individual transformation."[14] Almost as though he were acting out the concept, Sarton went on to adopt the explicit symbolic language of gift-giving, as he proceeded to mark and to facilitate my passage from apprentice to journeyman, my transformation from a graduate student into a junior member of the Harvard faculty. He proposed that I join a company which included the distinguished medieval historians Charles H. Haskins of Harvard, author of *Studies in the History of Medieval Science* (revised ed. 1927), and Lynn Thorndike of Columbia, well along on what would become his unique eight-volume work, *A History of Magic and Experimental Science* (1923-58), and, to go no further, the Yale neurophysiologist, bibliophile, and historian of medicine John F. Fulton, who had already published his magisterial bibliography of Boyle and would soon start work on his celebrated biography of Harvey Cushing. But that is not how

George Sarton phrased his invitation; here are his actual words:

3712.21

> Dear Merton,
> I have a fine proposition to make to you–as a Christmas present. Would you care to become associate editor of *Isis*, your domain being defined, e.g., [as] "social aspects of science"?–You would not be expected to do more for *Isis* than you have done thus far, but, I believe, this new title would be professionally helpful to you. Should you accept–as I hope you will–please send me as much of a "curriculum vitae" as you would like to publish in *Isis*. See vol. 27, 330.
>
> w.k.r.
> George Sarton

A year or so later, there is another threshold gift. To my mind, a gift of great symbolic magnitude; to Sarton's mind, evidently one also designed to advance my role as an academic journeyman. Here, in truncated form, is the letter carefully addressed to *Dr*. R.K. Merton:

May 3, 1939.

> Dear Merton,
> The fifth Congress of the Unity of Science will meet at Harvard University on Sept. 5-10, 1939. On one of these days not yet determined there will be a joint meeting of the International Institute for the Unity of Science and of the History of Science Society.
> In my capacity as chairman of the program committee of that special meeting I am now writing to you. The idea is to

have four items as follows [do note, once again, the company he asks me to keep]:

1. Prof. Werner Jaeger: Aristotle
2. Dr. De Lacy: Leibniz
3. Prof. G. de Santillana: The Encyclopaedists
4. Comte

I much hope that you will accept to deal with the last item. This would give you a good opportunity of distinguishing yourself . . . . The matter being urgent I would be grateful if you would answer it promptly, and much hope that your answer will be Yes.

With kind regards,
George Sarton

But alas, as I was compelled to report, I could not answer yes. For at the time of the Congress, I would be taking up my new post as contingent chairman of the Department of Sociology at Tulane University in the remote and inviting city of New Orleans.[15] In long retrospect, I think it is perhaps just as well for the relationship between my erstwhile mentor and myself that I could not accept the invitation to speak my mind on Comte and his positivistic descendants.

AN UNRULY APPRENTICE

That reflection concerning Comte gives me pause. I must not give the impression that all was sweetness and light between that mentor and me, that he was ever the benign, kindly spoken master and I ever the compliant apprentice. That was not the case. There were times, especially in his positivistic moments, when he was the exigent and angry master and I the brooding and unruly apprentice.

There was the time, for one instance, when I brought him, as a token gift in the asymmetrical reciprocities that mark the relationship of master and apprentice, an offprint of my first published paper-this appearing in a journal of sociology rather than, as I had hoped, in *Isis*. The gift was no doubt designed to intimate that my master's confidence in me was not entirely misplaced. Entitled "Recent French Sociology," it is as condensed and bibliographically crowded, if I may say so, as any entry in George Sarton's great *Introduction*. But in it I allude mockingly-not to say, flippantly and arrogantly-to "the enlightened Boojum of Positivism." My mentor did not take kindly to that facile (and Carrollesque) depiction. Still, this early episode led to little more than a symbolic rap on the knuckles. That was nothing at all to compare with my mentor's outrage, two years later, when I committed the cardinal sin of harshly criticizing Comtean positivism as set forth by F.S. Marvin (rather than criticizing that disagreeable man Comte himself, which would have been quite all right). That I should have done so as a guest lecturer at Sarton's invitation served only to compound the offense. That performance elicited this note:

3511.17

Dear Merton,

I think your talk was very good. Thanks. I was sorry to detect in your character a streak of cannibalism. At least your ferocious treatment of Marvin suggested that. Here is an old man who has devoted his whole life to the defense of generous ideas–you dismiss his collection of essays as if they deserve no attention. He repeats himself. Of course, he does; every one who has an important message *must* repeat himself time after time, for he knows that most people will only begin to understand at the 1000th time.

w.k.r.
George Sarton

Evidently I had touched an exposed Comtean nerve. And yet Sarton eventually did forgive–he was not of that Lethean mind that would forget-my behavior. Three years later, as I have reported, he was inviting me to speak on Comte at the exceptional joint meeting of the Unity of Science group and the History of Science Society. Thinking back on that earlier episode, I am inclined to agree with the import of Sarton's plain-spoken judgment on the style if not the substance of what must have been a blustering assault rather than a closely argued criticism. Perhaps I had engaged in the naïve and nasty game of simply scoring points off the other (absent) fellow, thus seeking to exhibit my seeming intellectual powers; and then again, perhaps not. I like to think that even as a callow graduate student I was well out of that unappetizing game. More in point,

I like to think that George Sarton's angry rebuke persuaded me once and for all that strong scholarly criticism need not be uncivil.

Along with learning from Sarton's response to the cardinal sin of pride (if not, I now say defensively, of sloth), I also learned from his response to venial sins in the scholarly life. I can tell here only a very few symptomatic episodes in the acquisition of craft skills and craft norms. From the start, Sarton had made it plain that it was good for a novice scholar to contribute his share to the common stock of knowledge through original research recorded in articles and monographs. But that was not enough. The role of the scholar called for more. One was obliged, for example, to do one's part in enlarging and facilitating the access of other scholars to the growing mass of knowledge claims by the writing of book reviews and bibliographic notes.

In this mode of scholarly work, Sarton himself was of course the incomparable exemplar. Deploying a typical piece of Sartonian arithmetic, an entry in his journal of 1952 estimates that, over a span of forty-one years, he had contributed about 100,000 notes to the critical bibliographies in *Isis*-those periodic, systematic, and annotated bibliographies which continue to this day. Sarton goes on to calculate: "I have written an average of *six* notes a day (holidays included). It is like the walking of 1000 miles in 1000 consecutive hours. To write six notes each day for a few days is nothing, but to do so without stop or weakness for 14,975 days is an achievement. It implies at least some constancy."[16] Nor did this calculus include the hundreds of his detailed reviews in the pages of *Isis*. Although he surely expected nothing of such magnitudes from others, he was concerned to set his novice on the right track. So it was during the half dozen or so years of my novitiate that I found myself writing, at a rather more restrained pace, some twenty articles and sixty-five reviews in various journals, along with those twenty entries for the critical bibliographies of *Isis*. Compared with the vast magnitudes sustained by my mentor, that seems little more than evidence of good intentions.

As editor-mentor, Sarton was also concerned to inculcate the norm of timeliness. After all, there were publishing deadlines to be met. Judging from his many handwritten notes to me on my reviews for *Isis*, he was generally satisfied with them on the counts of number, quality, and probity. But from time to time he was put off by my venial sin of procrastination. That sin he treated as a misdemeanor calling only for light, sometimes playful reproof. (That his sanctions were so gentle may help account for my recurrent attacks of procrastinitis over the years.) Thus, a note delivered to me at nearby Emerson Hall in 1937 consists simply of his temperate prod: "The 'scientist in action' [the title of a book by W.H. George] was sent to you last summer. What about the 'reviewer in action'?" That note is dated October 31st; the next note, postmarked the very next day, reads: "Many thanks for the very good review of

George's book." Evidently I had not long remained tardy.

In contrast to such mild injunctions for timely action, Sarton expressed intense commitment to another craft norm: the preparation of an adequate index to a scholarly book was for him a sacred trust. So it was that, in his reviews, he would severely rebuke the negligent authors of books that lacked an index or spotted one that was perfunctory and therefore largely functionless. Such authors were guilty of the moral dereliction of requiring serious readers who wanted to make limited, specific use of those books to engage in a drawn-out search through its pages and of requiring readers who vaguely remembered a salient passage in a book read some time before to reread much of the book in order to locate that passage. An index was for Sarton the instrumental expression of a technical norm and the symbolic expression of a moral norm, with the second supporting the first.

Along with the exercise of moral suasion and public sanctions, Sarton provided a prototype of indexing in unexampled detail. The first volume of his *Introduction*, running to almost 800 pages of text, has an index of 52 double-columned pages. The second volume of 1,138 pages of text has an index of 110 pages, supplemented by a "meager" Greek index of three pages. But it is the third volume (in two parts), with its almost 2,000 pages of text devoted to "science and learning in the fourteenth century," which engaged Sarton's indexing energies to

I

THE HISTORY OF SCIENCE AND THE HISTORY OF CIVILIZATION

ARE the main events of human history determined by a relatively few men or by the great masses of people? Are the so-called leaders really the leaders or are they led? Do they teach the people or are they simply its mouthpieces? Are they real creators or puppets? These questions have been discussed by each generation of historians, and two schools of historiography, which we might call the individualistic and the populistic, have held their own through the centuries. The individualists have naturally laid stress on biographies. From their point of view, a collection of biographies of the great men, of the "heroes" would be essentially the history of mankind. Their adversaries retort that selected biographies can never replace the history of the people itself, that they are at best but a part of it, and not the most important. It is clear enough that even the greatest general cannot win a victory without armies. Does he create the armies, or do the armies make him possible?

I suppose that one may go on arguing such questions until the end of time. The biographical method irrespective of its merits will always be very popular. From Plutarch down to Carlyle and to our own days, and not only in the West but in China and in Islām as well, it has always had its votaries and it has brought to light some of the masterprieces of historical literature.

After all, being men, we are primarily interested in men, and who represents them best, the conventional or unconventional heroes or their anonymous followers? We know well enough that men are exceedingly unequal in almost every respect but this very complexity increases our bewilderment. If they differed only in this or that particular, it would be so much easier; we could perhaps put them all in a row, starting from the worst at the extreme left to the best at the extreme right. But this cannot be done. Men are so different and they differ in such a multitude of ways that barring a few obvious cases, fair comparisons are almost impossible. The parents of Beethoven and Lincoln were plain enough, yet whatever influence was wielded by these two giants — their sons — must be traced back at least in part to themselves. In any scheme of society the parents of Beethovens and Lincolns are extremely important, but where shall we find them? In a vote we accept the judgment of the majority, not because the majority is necessarily right but because it is determined. There is but one majority while there may be any amount of minorities. In the same way, whichever be the real originality and power of the leaders, must we not select them simply because they alone are conspicuous, because they are more tangible and less undetermined?

I do not intend to enter into such discussions myself because I consider them unprofitable. In my opinion it does not matter so much upon whom the emphasis is laid, as upon what. Let the historian focus his tale upon a few individuals or upon many, I do not care. It is impossible to tell the whole of it anyway, and it is then largely a question of taste and art whether there are a few portraits upon the canvas, or many, or none but an anonymous and undifferentiated crowd. If the story is well told, the crowd must necessarily appear in one form or another, whether in the foreground, or in the middle ground, or in the background, — and if there be any concerted action, any directed movement, there must be leaders.

The leaders may be more or less self-willed and energetic, they may be put at the head or at the tail; the thing that interests me is the action itself, and its purpose or direction. My objection to many histories is not that they are too individualistic or the opposite, but rather that they are too frivolous. Much of the historical writing of the past, and not a little of the present, strikes me as being a sort of gossip, a sort of exalted gossip if you please. All of that pageantry centered around kings and solemn dignitaries was picturesque enough, but for one who really tried to understand the development of mankind highly irrelevant.

The main trouble with ancient historiography (by which I mean most of it until almost our own days) was not that it was focused upon too few individuals but that it was focused upon the wrong individuals. The early historian mistook kings for leaders and stewards for creators, he was more concerned with war than with peace and with disease than with health. His accounts were anecdotic and pathological; he paid more attention to the pomp of royal affairs, to the glamour of armies, to the vicissitudes of life in high places, to all the abnormalities and crimes of arbitrary élites, than to the quiet and unobtrusive activities of craftsmen, thinkers, and scholars.

But as soon as attention is directed upon the deeds which are truly constructive the difference between biographical and non-biographical history dwindles almost to insignificance. One historian may speak more lengthily of the architect of a cathedral, another, of the artisans, another, of the social conditions which made the undertaking possible or which in turn speeded or jeopardized it, — the essential is to explain how the cathedral was conceived and how it grew up. It was brought into being by the concerted efforts of many men and by the junction of many circumstances, but the main thing was its creation. As long as we account for that we cannot be far wrong. In fact the individual builders of many cathedrals are unknown; we admire their achievements as much as if we could name them, but our admiration is tinged with melancholy. Being what we are, imperfect and weak, the cathedral itself does not entirely satisfy us, we crave for intimate knowledge about its builders, we wish we could express our gratitude to them more personally. Yet even then we realize that the cathedral itself is the supreme object of our interest, even as it is the best memorial of its creators.

Before considering the very complex case of mankind as a whole, suppose we had to tell the history of a single man. How would we set about it? The main point of the story, I take it, would be to explain the development of his genius, the gradual accomplishment of his special mission. If he became a great mathematician, we would try to show how and when his mathematical bent revealed itself, how a growing boy devoted more and more attention to mathematics, how other interests were by degrees sacrificed to this dominating one. A boy who toys with mathematical ideas, what fun; but little by little they engross the whole of his mind until finally we have the awful feeling that there is no choice or freedom left. No more playing with mathematics, but rather mathematics playing with a human mind and using it to the limit. That is how genius looks when we come nearer to it. Nothing very comfortable or pleasant, but rather a fearsome mystery. Our story should be focused upon that very mystery. Its value will depend upon our ability to evoke the genius — everything else however much there may be of it being subordinated to this — to evoke its growth, its struggles, its fulfillment, its influence; it will depend also upon our success in making other people realize the mystery involved. It is clear that all else is relatively indifferent, in as much as we are interested in this man because of his mathematical genius. To be sure our curiosity is not restricted to the mathematical side of him — if we are sufficiently interested in his genius our curiosity is properly insatiable — but that side is the essential, every other, auxiliary. A biography which would be focused, let us say, on the account of his diseases, or of his loves or hatreds, might be entertaining, it might obtain the favor of superficial readers, but it would be false.

The case of mankind is not essentially different from that of a single man, though it is infinitely more complex. To begin with, the main direction is not so easy to discover, for there are many. Which is the purpose of mankind? Is such a question too ambitious? Is it at all possible to answer it? I believe it is. Without venturing into metaphysics, we may safely assume that the main purpose of any creature is indicated by its specific function. What can man do which other animals cannot? His purely physiological functions he shares with many of them; it cannot be that he lives only to live and reproduce his kind. Indeed if we look back we see that the men who came before us have not simply perpetuated their own flesh, but produced a quantity of things, material and immaterial, which constitute the best part of our inheritance. The totality of these things we call civilization. They include such material objects as buildings, statues, paintings, furniture, instruments and tools of every description, and such immaterial things as artistic and scientific methods, ideals, hopes, fears and prejudices. They represent the creative activity of man, his net creations above and beyond those which had no aim but to make his life possible, or to lighten it, make it more agreeable, and insure its prosperity and continuation. Is it not as clear as daylight that if we want to write the history of man it is this creative activity, specific to him, which must provide us with our Leitmotiv? Everything which pertains to that activity must be in the foreground of our picture; everything else, however interesting, in the background.

To put it briefly we might say that, as far as we can discern, the main purpose of man is to create such intangible values as beauty, justice, truth. I trust that the reader will not require any definition of these terms; that he can distinguish order from chaos, beauty from ugliness, justice from injustice, truth from untruth. It is not necessary that he be able to distinguish them in each and every case; there will always be enough ambiguous cases to rejoice the hearts of casuists, but we shall not allow the latter to sidetrack us. It will suffice to recognize that there have been in all times at least some men who were obsessed by the idea of creating beautiful things, of improving social conditions, of discovering and publishing the truth. The fact that they were not free from illusions, that their experiments were not always successful, that even the best of them made mistakes, does not affect the general statement. Considered as a body these men were those who fulfilled the distinctive mission of mankind, and to them we owe most of the privileges and of the pleasures of our lives, the nobility of our minds, the grace of our hearts.

Now these creative activities are of many different kinds, so different indeed that the men engaged in them may seem to walk along divergent paths. The artist, the social reformer, the saint, the scientist represent four distinct types, which may be occasionally united in various ways, but are generally separated. It would be foolish to attempt to dispose of them in a hierarchy. No one can say that this or that type of activity is necessarily superior to that other, for in every case it is less the type that counts than the manner. However for practical reasons we are obliged to single out and to put in the very center of the foreground one of these four main activities, that of the scientist.

Indeed the scientific activity is the only one which is obviously and undoubtedly cumulative and progressive. When we write the biography of a man we take pains above all to describe the development of his genius, the progress of his work. It is that very progress which gives its point to the story. In the same way human history is not truly significant unless we can describe the progress of mankind along a certain direction. But is there any real progress? It is characteristic enough that the old-fashioned humanists — of the literary or anti-scientific type — have often asked themselves that question, and have been unable to answer it. From their point of view the reality of progress is very doubtful. Are our saints more saintly than those of old, do they come nearer to God? Man does not seem to have succeeded in improving his sanctity, or for that matter, his wickedness. And our artists, are they getting any nearer to their goal of beauty? It does not seem so either. If Æschylus and Sophocles could attend our modern plays what would they think of them? What would they say of our art exhibitions? I imagine that their most charitable attitude would be to treat much of our effort not as a genuine performance but as a joke, — a huge and nonsensical joke. The fact is there cannot be any continuous progress in art or in literature. When one reads the history of science one has the exhilarating feeling of climbing a big mountain. The history of art gives one an altogether different impression. It is not at all like the ascension of a mountain, always upward whichever the direction of one's path; it is rather like a leisurely journey across a hilly country. One climbs up to the top of this hill or that, then down into another valley, perhaps a deeper one than any other, then up the next hill, and so forth and so on. An erratic succession of climaxes and anticlimaxes the amplitude of which cannot be predicted. This history makes one think of a rhythmic motion, or rather of many rhythms capriciously interwoven. For example, our artistic sensibility passes periodically from romanticism to classicism, or else from naturalism to idealism. There is no reason to change the direction of the movement except that the pendulum has gone as far up this way as it could and must come down again and up the other way. People get tired of romanticism, or idealism, or intense colors, or short skirts, or what not, and they want a change; sooner or later a point is reached where no change is possible except by reversing the movement. Under such circumstances there are only ups and downs; one cannot speak of progress, nor even conceive it. The very sophistry of those "humanists" is but a component part of the general rhythm; it is neither more nor less persuasive than that of all the sophists who came before them. Forsooth, the chances are that the modern sophists who parade under the banner of humanism would have been easily outdone by the Greek rhetors or by the mediæval schoolmen. As they have only a distorted knowledge of science and can see it only from the worst angle, as a purely utilitarian and materialistic activity, they have no compunction in discrediting scientific advances and exposing their futility. They will glibly say: "What is the good of being able to move twenty times faster than before if we have no place to go to? Of producing a hundredfold more goods if we cannot use them except for our own undoing? Machines have increased the quantity of things and destroyed their quality; they have created everywhere an artificial and senseless agitation, deafening noises and horrible smells; they have ruined one fair landscape after another and corrupted the countryside; they are responsible for all the horrors and iniquities due to the excessive human concentration of our octopean cities; they have poisoned forever the joy and innocence of men, and made it almost impossible to lead a quiet and meditative life."

We shall come back to the machines presently; they are only the by-products of man's scientific efforts. For the purpose of these efforts is not at all to increase speed, or to produce more goods than are needed; or to do any of the innumerable ugly things which they are accused of. The real purpose is to understand more deeply and more fully the whole of nature, including ourselves and our relations to it. An intense curiosity to find the truth about things in general and himself in particular is as much a characteristic of man as his thirst for beauty and justice. But it so happened that in proportion as he discovered nature's secrets, he was also able to make use of them for his own benefit, and in proportion as he disentangled its forces he contrived to capture and divert them to the gratification of his own needs. His main impulse was disinterested curiosity, but he found — that is, owing to the very consistency of nature he could not help finding — the magic formula, the "open sesame" which enabled him to tap the inexhaustible treasures of the earth and made of him the lord of creation. It is because of the limitless fertility of scientific efforts, of their enormous utilitarian and financial value, that the old humanists could hardly conceive they had any other value. In fact, though many scientific discoveries have created new power and wealth beyond the wildest fancies of the Arabian Nights, the greatest number have no practical value whatever. For the true scientist, these are not less precious. For him the infinite treasures which science has yielded and is constantly yielding are incidental; the main purpose of science, and its main reward is the discovery of truth. How very precious this discovery must be in his eyes if the unlimited might and wealth which science produces is comparatively of little account, — a by-product! But so it is. No scientist worth his salt would hesitate a single moment on this point, for he knows well enough that the discovery of truth is more valuable than any treasure. It is very similar to the discovery or creation of beauty, the reward being the same in both cases, namely that of contemplating quietly something which pleases the soul.

Let us suppose that Greek studies led occasionally to the decipherment of secret writings revealing the location of great treasures, would it be fair to say that Hellenists were nothing but treasure hunters, materialists lusting after gold and power? The attitude of many of the old humanists toward science is hardly more intelligent and more generous. It is as if their minds were obsessed by the gigantic prizes which scientific research has brought to some fortunate inventors. And of course how could they forget them when every day's newspaper contains sensational reports of new discoveries and more startling examples of the magical fertility of science?

It is said that at a scientific dinner in Cambridge, the main toast was proposed "To pure mathematics, and may it never be of any use to anybody!" This toast was of course a joke, but the joke would have had no point if it had not contained some truth. It expressed the impatience of many scientists with the excessive weight attached by the public to the utilitarian value of science. Many artists do manifest the same impatience when they hear people discuss the cost of works of art, because they well know that a thing of beauty, even as a clean particle of truth, is beyond price. It would be stupid to despise science because of its practical value; we ought to be grateful for it, we could hardly be grateful enough, and by the way, its blessings are not restricted to the scientists who discover them, but are shared by them with every man, with each one according to his intelligence or according to his need. Yet we do love truth just as much in the many cases where it has no practical or commercial value, where it has no power, except perhaps that of destroying our own prejudices and privileges.

Now as opposed to beauty, knowledge is cumulative and progressive. Looking at works of art will hardly help us to create better ones, but we can assimilate the store of knowledge amassed by the people who came before us, repeating in a few years the evolution of centuries, and start our own investigations where they left off. It is in that sense that one must understand the saying ascribed to one of the most lovable scholars of the twelfth century, Bernard of Chartres, "In comparison with the ancients we are like dwarfs sitting on the shoulders of giants." [1] Indeed from the point of view of science mankind might be compared to a single man, to a single giant whose knowledge and experience are steadily increasing.

But then is it not clear that if we wish to tell the history of mankind — the history of that giant — we ought to proceed as for a biography and focus our narrative upon the progressive elements, not upon the others? From the point of view of the historian of science, the growth of that giant, the development of his memory, intelligence, and power, are unmistakable and relatively simple; one can give a very full account of it. On the contrary the development of his artistic and religious possibilities is far less obvious, and may even be doubted.

However at the risk of weakening my own argu-

[1] From the Metalogicon of John of Salisbury, Bernard's pupil (Book 4, Chapter 3). The exact quotation reads: "Dicebat Bernardus Carnotensis nos esse quasi nanos gigantium humeris insidentes, ut possimus plura et remotiora videre, non utique proprii visus acumine aut eminentia corporis, sed quia in altum subvehimur et extollimur magnitudine gigantea." A similar saying is often ascribed to Newton. See *Isis* (**24**, 107–09; **25**, 451; **26**, 147–49).

ment, I would not deny the reality of progress in the non-scientific fields; it is far less tangible in those fields but not inexistent. To be sure our artists are not greater than those of the golden ages of Greece and China, to be sure we do not produce more beauty or beauty of a higher kind, but can it be denied that whatever beauty there is may be enjoyed by a larger proportion of the people? The old civilizations were based upon the existence of slaves or their equivalent; only very few men were allowed to share its blessings, and needless to say, this does not mean that every one of these few privileged men did actually share them. We must assume that then as now there was a vast difference between the material and the psychological ability to enjoy beauty. For example, then even as now it was not enough to own a beautiful pot in order to appreciate the symmetry and the elegance of its shape. At present on the contrary many of the purest artistic joys can be shared by multitudes, each individual partaking of it according to his own merit. Think of our museums where hundreds of beautiful objects are collected and exhibited with loving care and touching ingenuity for the benefit of anybody who deserves it not because of his station in life, but only because of his own virtue. Is not all of this a real progress from the point of view of beauty? Of course we know that there are connoisseurs who do not fully enjoy a thing unless it be sufficiently exclusive; their love is jealous and mean, but this is obviously a perversion. We should feel, and most of us do feel, that our enjoyment of beautiful things is not impaired by its being shared with others but on the contrary multiplied. My own joy at a concert is greatly intensified by the consciousness that so many people are sitting around me who experience the same emotion; I could not bear to sit there alone. Now such sharing is more and more the rule of modern life. There may not be more beauty, but whatever there is, is as if it were infinitely multiplied by the number of hearts which are enabled to partake of it.

There may still be slaves and there are certainly a great many drudges even in the most civilized countries, but their number is steadily decreasing and the possibilities of emancipation are more numerous. There is no hopeless slavery except that which is inherent in a man's own abjection. The slaves have been replaced by machines, and if the latter have often been abused, their inventors cannot be blamed for it but only the selfish and accursed men whose immoderate greed turned blessings into nightmares. Whenever this happens (and it has happened but too often) we know that it is only a temporary error, frightful enough as long as it lasts, yet remediable. Modern civilization (this "machine age") is essentially different from earlier ones because our knowledge of the world is deeper, more precise and more certain, because we have gradually learned to disentangle the forces of nature, and because we have contrived by strict obedience to their laws to capture

as soon as it is completed the contradictions will disappear.

Scientific facts might be compared to numbered links which are joined together in their numerical order and form chains of various lengths. Ultimately they will all be linked, and in more than one way, and it will happen that whatever the linkages, the order of the links will not be affected. But as our knowledge is still very far from perfection many sets of links are still separated from the others. Sooner or later the missing links will be found, and we know that they will fit, that is, if they be the right ones, for such adjustments are very frequent. The chains of knowledge are not built in the simplest manner, but in a round-about way "as the wind bloweth." When we say that science is essentially progressive this does not mean that in his quest of truth man follows always the shortest path. Far from it, he beats about the bush, does not find what he is looking for but finds something else, retraces his steps, loses himself in various detours, and finally after many wanderings touches the goal. It takes him much longer to reach the aim but his knowledge is incomparably richer when he does reach it. None of these accidents influences the final results; the links and the chains are entirely independent of the capricious order of discovery.

Now the search for the truth is not restricted to any single group or class or nation of men. If one takes the whole of the past into account, not simply one period, and all the chains, not simply a few of them, one finds that men of all kinds have shared in the work. No one can predict where and when the missing links of any chain will be discovered, and these links will be entirely independent of the discoverers. This proves that with regard to this, its highest task, mankind is deeply united.

The political historian who is obliged to devote considerable attention to all the differences and jealousies which break mankind into a number of antagonistic fragments, does not realize that deep and secret unity. He is used to think in terms of competitions and stresses, and the conflicts between nations and the open hostilities they lead to are of course far more conspicuous than their common aspirations and obligations. He does not begin to realize that no matter how inimical one nation may be to another, or one class of people to the other classes, as soon as it comes to science they are indissolubly bound together. As soon as they search for knowledge they must needs follow the same path, at least part of the way; whether they wish it or not, they cannot help collaborating.

The unity of science and the unity of mankind are but two aspects of the same truth. Take it as you please, it represents the central direction of human thought. We do not know whither mankind is bound, we do not know the final goal, we cannot apprehend it for the simple reason that we are too far from it, but we know the general direction, and we know with at least five millennia of experience to depend upon, that that general direction, which is determined by our scientific efforts, is essentially stable.

These two complementary ideas suggest the dualism already referred to, and may lead to two different conceptions of the history of science. One may insist on knowledge itself, and write a history which being essentially a history of ideas, is very abstract, or one may insist upon the human side, the capricious origin and development of discoveries, all the little accidents which set off our curiosity in all kinds of directions, and obliged us to turn around the goal in ever narrowing circles before we finally touched it or came sufficiently near to it to be clearly conscious of it. The complete historian should join both tendencies. He should keep in mind and use as his guide the concatenation of pure ideas, such as can be reconstructed when all the mistakes have been made and corrected, but he should never forget the very humble origins of our boldest theories and their abundant vicissitudes. The abstract type of history may be very instructive from a technical or philosophical point of view, but it is deeply misleading, because it gives us an impression of simplicity and directness which is as unreal as anything can be. The scientific travail of mankind was never easy, never simple, and the fine abstractions it produced were always mingled with a large quantity of concrete facts and irrational thoughts from which they had to be painfully extracted.

The prime mobile of scientific progress was man's curiosity, a curiosity which was too deep-rooted to be interested in the ordinary sense, or even to be prudent. This is admirably symbolized by the tale of the tree of the knowledge of good and evil which stood in the midst of the Garden of Eden. Adam was forbidden to eat of it, but the serpent tempted Eve, and she tempted her husband and they did eat. Their eyes were opened, they lost their innocence, and the endless quest of truth was irrevocably begun. Time after time that story has been enacted again. Men have been forbidden to eat of other trees of knowledge, but they have finally eaten of them; nor could they help it. Once that it had been aroused, there was no way of stilling their hunger for knowledge.

But aside from that primary cause, there were many others. It would not be too much to say that the progress of science is a function of every human activity, of every human passion, good or evil. This might be illustrated by the history of geography. We know of a number of explorers who drew enough courage to brave terrible dangers and mysteries more terrible still from their scientific curiosity and from their love of glory. But we also know that most geographical discoveries were made incidentally by men who cared less for knowledge than for power, and less for glory than for wealth. Others were due to the ambition of kings and conquerors, to their rivalries, to their greed for gold, or spices, or slaves, sometimes to their desire to convert the heathen and to extend Christ's empire together with their own. And how many sprang out of the love of sport and adventure? How many of the explorers ran away from their home country which for some reason had become too hot for them, how much of their effort was due to such repulsive forces rather than to attractive ones? It is impossible to fathom the complexities of human hearts; it is impossible to judge. Those who seemed the most disinterested were perhaps more interested than we imagine, and vice versa.

The history of inventions would lead to similar conclusions. Some inventors died in poverty, others amassed considerable wealth, but it does not follow that the latter were more greedy than the former. In fact a successful inventor may be more disinterested at heart than a pure mathematician whose work will never have any commercial value. It is well to remember in the first place that even those inventors whose worldly success was the greatest could not enrich themselves without enriching mankind considerably more. In the second place, that their activity may have been inspired by internal or external factors bearing no immediate relation to the results, e.g., the love of women, the needs of different industries, prohibitive taxations, wars and blockades. Some men did their best work under the pressure of necessity and their minds seemed dominated by outside events, others created their own necessities irrespective of circumstances; some men were goaded by poverty, others would have been paralyzed by it.

But in a general way we shall not be far wrong if we assume that the central urge was largely unconscious and instinctive; it was due to the presence of the very qualities which were needed, and first of all, that irrepressible curiosity of which we have already spoken. Why does a boy become a musician? Because he is a musician and always was, because he was bent that way in his mother's womb. Why does another become an inventor? Because he was born that way. It was simply in every case a natural development of latent possibilities. To that extent, no matter what they did, their activity was disinterested. In a higher sense we may assume that the purely creative activity is always disinterested, if not in the initial stage, at least later on when it is thoroughly heated up. A man may dream of an invention which will bring comfort to himself and his family; to become rich may seem to be his main stimulant; yet as he continues his research and becomes more and more engrossed with his schemes and devices he may forget his own interest and even lose the deep-rooted instinct of self-preservation. He may finally reach that stage of spiritual intoxication, of complete self-forgetfulness which is perhaps our nearest approach to heaven.

This complexity of feelings may be illustrated by

the example of Charles Goodyear, who invented the vulcanization of rubber and thus became one of the great benefactors of mankind. He made many other discoveries relative to rubber manufactures. He tried hard all his life to enrich himself, but he only succeeded in enriching others; he died a poor man. I do not believe that his inventions were disinterested, yet by the end of his life he had become almost indifferent to money, and he made the following statement, the simplicity and candor of which impress me very much: "The writer is not disposed to repine and say that he has planted and others have gathered the fruits. . . . Man has just cause for regret when he sows and no one reaps." [1]

Mankind's purpose is served almost equally well by the greed of some men and the disinterestedness of others, and what each inventor got for himself out of his own discoveries or failed to get is after all a secondary matter, and almost irrelevant. However great the financial and other material rewards may be they are insignificant as compared with the purely spiritual ones, the feeling of having done well, and above all the pure contemplation of truth.

This had been clearly seen by the Greeks, witness this beautiful fragment of Euripides:

"Blessed is he who has attained scientific knowledge, who seeks neither the troubles of citizenship nor rushes into unjust deeds, but contemplates the age-

[1] Quoted by Holland Thompson: *The Age of Invention* (New Haven, 1921, p. 174).

less order of immortal nature, how it is constituted and when and why. . . ."

I hope I have succeeded in showing that the scientific activity, however abstract its fruits may be, is nevertheless essentially and intensely human. But if it is so human, and withal so important, how is it possible that historians have paid so little attention to it and that the old-fashioned "humanists" have even affected to ignore it altogether and to consider it irrelevant to their purpose?

The explanation is simple enough. That activity is to a large extent inconspicuous and even secret. It is impossible not to see the soldiers marching to war, nor to hear the blowing of the trumpets and the noise of the battle; it is impossible not to see the king sitting on his golden throne, the regal and holy processions, bishops blessing the multitude, and many other grand spectacles, which seem to symbolize the whole of life and the best of it. But how many of us will manage to see the artist painting in his studio, the scientist meditating in his garret? The case of the latter is extreme. For the artist's canvases will finally be exposed to public view, or his music will be played

[1] Ὄλβιος ὅστις τῆς ἱστορίας
ἔσχε μάθησιν,
μήτε πολιτῶν ἐπὶ πημοσύνην
μήτ' εἰς ἀδίκους πράξεις ὁρμῶν
ἀλλ' ἀθανάτου καθορῶν φύσεως
κόσμον ἀγήρων, πῇ τε συνέστη
καὶ ὅπῃ καὶ ὅπως.

A. Nauck: *Tragicorum graecorum fragmenta* (2nd ed., 1889, no. 910).

and make many a heart quiver a little faster, but how many people will ever understand what the scientist meant and did? It is not only his activity which is secret but even the products of it. Sometimes he may appear for a moment in the limelight, but on the whole these occasions are rare and if he be a good man as well as a good scientist he will not wish to multiply them. They are still rare to-day, they were considerably more so in the past. When scientists are publicly praised, it is more often than not for secondary and inferior achievements.

This is a paradoxical situation. The most conspicuous activities of mankind are comparatively insignificant as far as the attainment of its main purpose is concerned; the most important activities, those which are essential to that purpose, are hidden. Hence we may say that the history of mankind is largely secret. The results of mankind's fundamental efforts appear from time to time at the surface, but the long and complicated process of obtaining them is unsuspected except by a very few. But is this so strange after all? And is not the case of mankind in this respect very much like that of a single man? For which of our activities are the most manifest? A whole crowd may see us eat in a restaurant, or walk in the street, or they may hear us talk, but our real work — can anybody but ourselves be aware of it? One may have the illusion that one sees a man working, and this is truly possible for the lowest kind of work, but can we see him think? We may watch a physicist in his laboratory, but this will not help us very much. When he seems most busy, it is probable that he does nothing of importance. For aught we know he may be doing his best work while he is shaving or playing with a little dog. This explains the disappointment of the good people who go and disturb a great man with the naïve hope that they will see something; and of course they see very little. They meet a man who may be kind to them, but the real man, him whom they come to see, is not there at all. He is awaiting their departure to be himself again.

And so it is with mankind. Two great events happened in the year 1686, the publication of Newton's *Principia* and the constitution of the League of Augsburg. Everybody discussed the latter, but only a relatively small group of men were at once (or ever) aware of the former. The political importance of the League can hardly be exaggerated, but after all the world wherein we are living to-day, would not have been essentially different if the League had not been brought into being. As to the *Principia*, this was really the foundation stone of modern thought. Our conception of the world was utterly changed by it; that is, the world itself was changed. There are thousands of professional historians, but how many of them see those two events in their true perspective? Very few. In fact for most of them the *Principia* are practically inexistent.

There comes back to my mind very often, when I think of that, the old Heraclitean saying, "The secret harmony is better than the visible one."[1] The secret harmony is that which science reveals to us, all the beautiful and complicated symmetries of the cosmos, all the rhythms which our differential equations outline with such elegant brevity, all the graceful details of structure and function which scientific research in almost any sphere brings to light day after day in endless abundance. But it is also, and it is chiefly of this that old Heraclitus makes me think, the secret development of man's destiny. The visible activities of man are so many, and some of them so brilliant, so spectacular, so pleasant to look at, — yet his main activity remains secret. A spectator who sees only the surface of things, however much he may enjoy and admire them, must finally ask himself: "What is it all about?" Man seems to be turning in a hopeless circle. Yet underneath that fantastic agitation there is going on all the time a slow but steady process of creation. The majority of men are hardly aware of it while it lasts; but they are ready enough to take pride in some of its fruits later on. They themselves recognize among the greatest men of the past, the artists, the poets, the saints, and sometimes the scientists, who were indeed the main actors. They realize more or less consciously that it was these men who accomplished the destiny of the race. However in

[1] *Ἁρμονίη ἀφανὴς φανερῆς κρείττων.* H. Diels: *Fragmente der Vorsokratiker* (2nd ed. of vol. 1, Berlin, 1906, p. 69, no. 54). The medal which was struck a few years ago to the memory of the great mathematician Henri Poincaré, bears very appropriately that very legend (see *Isis*, vol. 9, p. 420, pl. 19).

most cases they do not recognize such activity until it has been halted by death. And so it is for example that the personality of one of the greatest dramatists of all times has remained strangely elusive. We know in great detail the lives of a number of Elizabethans who were famous in their days but the life of William Shakespeare is so badly remembered that it has been possible to ascribe his works to various contemporaries of his, that is, to suppress him altogether. To be sure these attempts have failed, but the very fact that they were made at all proves our ignorance. People knew well enough the men who "did things," — and Shakespeare was not "doing" anything, was he? Within three centuries public judgment has changed enormously, and which do you think is the right judgment, that of the contemporaries who considered great and important hundreds of nobodies, or that of posterity? After all, posterity is impartial, it is not deceived by outside appearances, it has plenty of time to reach its conclusions. I have taken Shakespeare as an example because it is the most striking, one of which anybody can appreciate the cogency, and because in terms of human history, he is very close to us; one cannot put the blame on the distant past, the "dark ages." The brutal fact is that one of the greatest poets of all ages was living in England not very long ago, and so few people were aware of his greatness that his personality was never allowed to emerge into the light. In the meanwhile this poet by his own unaided efforts was raising the

English language and the English genius to a much higher level. He was building up England but England knew him not. Is this not secret history? As to the scientists, our ignorance of them is far greater; most of them are entirely unknown even to very educated people. For example, how many Elizabethan scientists do you know, and how much do you know of them?

There is still another reason why the history of science is not as popular as it should be, and why the great scientists of the past have not received the same tribute of homage as, let us say, the great artists. The majority of men if they but enjoy a modicum of comfort are deeply conservative and afraid of change. Now scientific curiosity being the main source of progress, is also of necessity the main cause of change in the world; in that sense, it is the most revolutionary activity of our mind. Its revolutionary tendency is not restricted to this or that, it extends to everything. The scientific spirit is never at rest. It is never blindly satisfied with what is, it wants to improve it if possible, or to replace it by something better. It is always preparing new experiments into the unknown; it is essentially adventurous. Most people have the obscure feeling that the scientist is the great trouble maker and joy killer. Is he not always urging them to advance, when they would prefer to rest, and to take additional pains when they would say: Let well enough alone? Moreover knowledge might be compared to the sun whose rays destroy microbes wherever they fall; individual and social diseases flourish better in the darkness. Throw the light of knowledge upon them and sooner or later they must wither. Ignorance and injustice are driven out in the same fashion. No wonder that all those who enjoy undeserved privileges and are afraid of losing them fear the curiosity of men of science. And then again there are all the superstitious relics of the past to which people hold as passionately as mothers do to their less favored children. These superstitions may be picturesque, they may have some lovely aspects, but they are always dangerous, not simply in themselves, but because of all the falsities and miseries which are likely to lurk behind them. The scientist finds in his heart no pity for them; he cannot tolerate them any more than weeds in his garden or parasites on his body. They must go, and it happens that harmless things, things that were not unlovely, are uprooted together with the weeds and thrown with them upon the dunghill. This causes many good people to suffer and to resent what they might call the meddlesomeness of scientists.

Science tends to destroy the darkness where evil and injustice breed, but there is also some element of beauty and poetry in that darkness. The most perfect kind of beauty is not afraid of light, but perfection is rare. A pretty girl in the bloom of her age may glory in the sunshine; a middle-aged woman prefers a subdued light. In the same way there are many things in life which are still beautiful but not enough to stand the full glare of the sun, and when the scientist insists upon turning his pitiless searchlights upon them, he hurts them and scandalizes faithful hearts.

It is well to admit that much, but we must remember that it is partly unavoidable. And it is certainly not true to say that science destroys poetry and mystery. It destroys some of it, but the little which is lost is richly compensated by the revelation of the infinite beauty of the hidden world. Remember! "The hidden harmony is better than the visible one." Science is all the time analyzing and pulling down little mysteries, or rather displacing them. As the known world increases, its frontiers with the unknown lengthen and the mystery deepens. There is far more mystery in the universe for the learned than for the ignorant but it is otherwise distributed. The purpose of science is to make clear distinctions between what we know and what we don't know, and furthermore to allow us in every case to measure the degree or the quality of our knowledge. Mysteries which we have driven outside of the boundaries of our knowledge and which we have located and encompassed, such mysteries will not harm us; on the contrary they will stimulate and inspire us in many ways; the dangerous mysteries are those which are hopelessly mingled with our knowledge, and of which we are perhaps unaware. The scientist will never relent in destroying the harmful mystery, but the sum total of mystery and poetry cannot but grow together with the world which his adventurous mind is reflecting.

This brings us back to the very individualistic nature of the history of science, or of the history of civilization focused upon it. For civilization has been created by relatively few men in the face of the vast majority of their brethren. Each step forward has meant a protracted struggle against the fears and prejudices of the crowd. When I speak of the crowd in this connection, I do not mean the inarticulate mass of the poorer people; on the contrary the "crowd" resisting progress includes men of all classes, rich as well as poor, the most powerful as well as the most forlorn; it even includes many of the conventional leaders, kings and great stewards, popular preachers, moulders of public opinion. No novelty, whether in art, or in religion, or in science could be really introduced as long as the hostility open or latent of the people had not been overcome. If the contest was short and weak, the chances are that the novelty was not worthwhile or that it was purely superficial.

Popular resistance has been especially strong against religious reformers and men of science. That saints and scientists should thus be placed in a single category by the opposition of others is far more than a casual occurrence. They have much in common, above all, their otherworldliness. There are but few saints among scientists, as among other men, but truth itself is a goal comparable to sanctity. As the Pythagoreans had already understood it more than twenty-four centuries ago, there is sanctity in pure knowledge, as there is in pure beauty, and the disinterested quest of truth is perhaps the greatest purification.

However the hostility to saints and to investigators was not altogether of the same kind. Attempts at religious reform aroused popular anger because the inborn conservativeness of man is nowhere stronger than in the field of religion. The religion of his fathers must not be criticized even if his own profession of it is but an outward show. The most malicious kind of hatred is that which is built upon a theological foundation. On the other hand the resistance to scientific novelties was due to an intuitive, if unconscious, appreciation of their revolutionary nature. The slightest and the most innocent scientific innovation is but a wedge which is bound to penetrate deeper and deeper and the advance of which will soon be impossible to resist. Conservative people are undoubtedly right in their distrust and hatred of science, for the scientific spirit is the very spirit of innovation and adventure, — the most reckless kind of adventure into the unknown. And such is its aggressive strength that its revolutionary activity can neither be restrained, nor restricted within its own field. Sooner or later it will go out to conquer other fields and to throw floods of light into all the dark places where superstition and injustice are still rampant.

The scientific spirit is the greatest force for construction but also for destruction.

One cannot properly understand the history of science, or indirectly the history of civilization, unless these struggles, which are like the growing pains of mankind, are carefully described, for the opposition to scientific advance is one of our means of measuring the latter. Moreover that opposition is itself not only useful but indispensable. The reality of progress implies some organized resistance to it. Without resistance there would be no stability, and hence no order and no gain, only chaos. Any form of antagonism puts men of science on their mettle and obliges them to be more careful and conscientious. One must not forget that the scientific advance is not always nor necessarily in the right direction. Every resistance is a welcome test. After all, mankind has a perfect right to distrust novelties and to exact abundant proofs of their value before accepting them.

The man of science walking audaciously ahead of the human procession is the great tempter. Which new ideas is he going to offer us next? Mankind would like to sit down and rest, but he must go ahead; no peace for him or with him, for he is its restless spirit, its very conscience. The saint and the artist are troublesome enough, for they hunger for sanctity, for beauty, for justice, and will never be satisfied as long as there is wickedness, ugliness and injustice in the world. The man of science is the most troublesome of all for it is not enough for him to improve what is, he wants to explore the unknown, full of mysteries and terrors; he cannot be happy as long as the unknown exists, and yet as he proceeds there is more and more of it. Thus is poor mankind endlessly dragged behind these tyrannical heroes and obliged by them to accomplish its appointed task against its own will. No wonder that it often hates them, and cannot fully honor them until they are safely dead.

It is not difficult for us to sympathize with poor bedraggled mankind, for the struggle of which I spoke exists also on a much humbler scale in every one of us. You see how my comparison of mankind with a single man verifies itself at every step of the argument. Each of us has a spirited and adventurous master who presses him to go ahead without fear and without cease; each of us alas! has also a weak body — our own brother ass! — which would rather take it as easy as possible and not try anything new or risky. The issue of the struggle is very different from one man to another. Sometimes the spirit is always triumphant; sometimes brother ass is always the stronger, and the spirit is all but extinguished. In most cases there are endless vicissitudes, ups and downs, and we may be very spiritual one day, and very carnal and sluggish the next one.

The realization of this inward struggle should help us to understand the greater struggle which has been going on for ages and will continue indefinitely between a few spirited leaders on the one hand and the heavy and lazy bulk of mankind on the other. It should help us to recognize that the men of science are not troublemakers in any other sense than that our conscience is a troublemaker. That is, they make trouble but for our ultimate good. Without consciousness we should slide back to the level of brute beasts. Without scientists, without saints, without artists, mankind would soon be reduced to a society of animals. Without saints, it would fall into sin; without artists, into ugliness; without scientists it would stop altogether and degenerate.

It is very well to be conscientious, and we cannot be too conscientious, but it is a thousand pities to be self-righteous. I am afraid that scientists have sometimes shown a deplorable tendency to be too proud and assertive, and — as a class — to evidence too much self-righteousness. Some of them have been foolishly aggressive toward all non-scientific activities and have created more antagonism against themselves than would have been otherwise the case. Others have behaved like intoxicated boys, destroying ruthlessly every thing which seemed wrong or irrational in their own eyes, and have proved themselves reckless iconoclasts, more stupid and less excusable than the superstitious image builders. Such mishaps are as deplorable as anything can be, but cannot be entirely avoided. The truth is, a man of science is not necessarily wise; his mind may be very acute and yet very narrow; he may be able to penetrate mysteries veiled to all other men, and be in that respect of almost uncanny intelligence, and yet be very dull and dense in every other. Finally it must be confessed that many men of science show a lack of education which cannot but irritate the people upon whom they are looking down and who are peradventure far more civilized than they are.

As time goes on such disharmonies are harder to bear. We do not believe any longer that a saint is more saintly for being unwashed and uncombed; we do not believe any longer that a scientist who behaves like a bull in a china closet is pardonable. "Noblesse oblige" applies to learning as well as to anything else. The better and deeper a man's knowledge, the higher his humanistic possibilities, the higher also his humanistic responsibilities. If it turns out that in spite of his learning he is after all but an uneducated and graceless man, so much the worse for him, so much the worse for his learning.

Precisely because he is what he is and because of the revolutionary possibilities of his investigations, the scientist should take special pains to know the past, that is, to know the history of science and the history of civilization as I have defined them. Inasmuch as the natural drift of his mind is forward — and perhaps dangerously so — he should try to understand the origin and development of his ideas, he should try to contemplate as often as possible the achievements of the men who came before him and to whom he owes everything that he possesses, everything that he knows, everything that he is. To be amiable the man of science needs if anything more historical knowledge than another man, not less.

The progressive outlook should always be combined and tempered with reverence. It is man's destiny to go forward, and to try to stop him on this, his essential errand, would be so childish and futile that it is not worthwhile to think of it; but he should always keep in touch, and reverently so, with his predecessors. Reverence without progressiveness may be stupid; progressiveness without reverence is foolish and wicked.

The study of the history of science would teach the scientist that he must be tolerant of others, because the ways of mankind are very uncertain at best. That is, man does not generally find the right path until after having groped around it for a considerable time and having lost himself in many detours and blind alleys. He does but seldom follow the shortest road from.one discovery to another, for the shortest distance to it can only be determined when the new discovery has been made. It is only retrospectively that the real direction of man's endeavors can be distinguished among all the parasitic ways in which he squandered his time and energy. The chances are that, imperfect and limited as he is, man will continue to beat about the bush and make one mistake after another in the future as well as in the past. Instead of judging severely the errors of his forefathers he should be grateful to them for having made them, and thus helped him to avoid them. He should never forget that if he is able to see further than they it is only because he is standing on their shoulders.

Scientific activity is the foremost creative activity of mankind, not only materially but also spiritually. Think of how the universe has been enlarged in every direction by the investigations of astronomers, physicists, and biologists. What a distance between the very small world of Genesis centered upon the Garden of Eden and the universe created by modern science. No poet ever dreamed dreams comparable to the realities unveiled by the scientist, and these realities do not preclude new dreams but force them up to a higher level. In their light petty dreams become more and more incongruous, while nobler ones are countenanced. Indeed the sphere of dreams grows together with the sphere of reality. Knowledge, not ignorance, is the cradle of true poetry. To be sure the creative activity of the scientist involves a certain amount of destruction. The greatest builders of the race should be allowed to destroy what must be destroyed, — as little as possible. Let them suppress the ugly things, the iniquities, the superstitions, the oppressive remains of the past, and those only; let them quench the nightmares, but not the adventurous dreams, not the pure poetry which is our projection into the future.

The only way of humanizing scientific labor is to inject into it a little of the historical spirit, the spirit of reverence for the past, — the spirit of reverence for every witness of good will throughout the ages. However abstract science may become, it is essentially human in its origin and growth. Each scientific result is a fruit of humanity, a proof of its virtue. The almost unconceivable immensity of the universe revealed by his own efforts does not dwarf man except in a purely physical way; it gives a deeper meaning to his life and thought. Each time that we understand the world a little better, we are also able to appreciate more keenly our relationship to it. There are no natural sciences as opposed to humanities; every branch of science or learning is just as natural or as humane as you make it. Show the deep human interest of science and the study of it becomes the best vehicle of humanism one could devise; exclude that interest, teach scientific knowledge only for the sake of information and professional instruction, and the study of it, however valuable from a purely technical point of view, loses all educational value. Without history, scientific knowledge may become culturally dangerous; combined with history, tempered with reverence, it will nourish the highest culture.

The most ominous conflict of our time is the difference of opinion, of outlook, between men of letters, historians, philosophers, the so-called humanists, on the one side, and scientists on the other. The gap cannot but increase because of the intolerance of both and of the fact that science is growing by leaps and bounds. The old humanists who claim that science has only a technical function and who would say to the scientists "Stick to your last, keep to your technicalities, spiritual matters are our business," — would widen the gap beyond the possibility of healing. For the happiness and the sanity of mankind let us hope that their plans will miscarry. The present situation, it should be noted, is only a beginning, not a climax. The bewildering abundance and the complexity of modern science are as nothing compared with those which will obtain a hundred or a thousand years hence when our modern science will be ancient. As science grows much faster than anything else, its relative importance in life must necessarily wax. What would happen if all the scientific knowledge and material power were concentrated in the hands of one group of men, and all the educative opportunities in those of another? Heaven forbid! The situation is undoubtedly aggravated by the aloofness of many scientists. That aloofness is not entirely their fault. It is caused by the overpowering combination of two forces, on one side the attraction of their own studies, their extreme concentration in the subjects which they are investigating, on the other side their repulsion by the old humanists, the feeling that their cooperation is not desired.

Instead of broadening deliberately the rift which separates them, would it not be wiser and gentler to bring these two sets of people closer to one another?

In place of the intolerant attitude of the old humanists I would suggest the opposite one. Humanism, that is, education, culture, is or ought to be the common good of mankind. Every creative activity in the right direction was, is, and will be a contribution to it. Humanism is not and cannot be the monopoly of any group of men; it is the ultimate result of all efforts to increase the intellectual value of life, it is the sum of all disinterested efforts from the humblest to the highest. As it is essentially an integration, it is clear enough that it can never be attained by expelling arbitrarily the most powerful group of creators, — of "poets." To complete the integration, each group must learn to understand the other. The educated people in general must obtain some knowledge and appreciation of science; the scientists must receive some historical training, must be taught to look backward as well as forward, and to look with reverence. These good offices may be rendered to both groups by the teaching of the history of science and of the history of civilization focused upon it, — the noblest part of our history, the one which leaves us neither shame nor regret.

Think of a young scholar who is trying to revive the spirit of Greece. How will he ever succeed in doing so if he fails to capture the spirit of his own times? The Athenian humanists took the whole of knowledge as their province; they did not capriciously exclude this or that; they deeply realized the unity of science. How will the young scholar understand their attitude if he has been brought up himself in ignorance of the most exuberant activity of his own days, and been taught to consider it as of no cultural interest, — a simple collection of technical and utilitarian recipes? And on the other hand think of a young physicist working quietly in his laboratory, undisturbed by all the noise and glamour of a feverish and whimsical world, working quietly as if he had the whole of eternity before him. Is he a humanist, or not? It all depends on his own education, on the quality of his own soul. The chances are however that all is not well with him. His aloofness is too great; he is too proud perhaps, too deeply engrossed in his own task to see it in its proper perspective; he may know well enough what he is doing, but is he sufficiently aware of the multifarious activities of his fellowmen?

The only cure of the first is to make him realize the real meaning, the cultural value of science, which far transcends all its applications, and this may be done best by teaching him its history. The only cure for the second is the contemplation of the past, the whole past, but especially the development of civilization as focused upon the very efforts of his predecessors. Let us lift up to their eyes the great mirror of history, and teach them to use it as a mirror not of their own good selves but of the whole of mankind.

Between the old humanist and the scientist, there is but one bridge, the history of science, and the construction of that bridge is the main cultural need of our time. An immense task to be sure but one worth every pain it may cost. I do not know who is the poorer: the old humanist without understanding of science, or the scientist without appreciation of beauty, without urbanity, without reverence. I do not know which is worse: idealism without knowledge, or knowledge without idealism. We need both equally in order to go forward and prepare the dawn of a new age, — the age of a New Humanism.

II

EAST AND WEST

WHEN one speaks of the history of science most people think of experimental and mathematical knowledge as we know it now, with its inexhaustible harvest of applications; they think of what we would call "modern science," the development of which was hardly started before the seventeenth century. This is of course justifiable in some respects, yet he who was acquainted only with that part of the story would have a very misleading idea of the whole evolution. It is as if he knew a man only in his maturity and was not aware that such maturity was made possible only by the long years of childhood and adolescence.

This recalls my comparison of mankind with a single man, and how it helped us to understand both. Let us return to it once more. What would you think of a biography which began, let us say, at a time when the hero was thirty, was married and already had children, and was well started on his work? Would not such a biography be very disappointing? For we would want to know how he got started, whom he had married, how he became interested in his chosen work and gradually devoted all of his thought and energy to it. For exactly the same reasons a history of science beginning only in the sixteenth or seventeenth century is not only incomplete but fundamentally wrong. This is even more true in the case of mankind than in that of a single man, because in the latter case we can at least imagine various possibilities. If we have read many biographies of men of science we have in our minds a sort of composite picture of their youth which may serve as a first approximation. But in the case of mankind it is simply impossible to imagine the history of the four or five millennia of recorded experience which preceded the advent of modern science.

It is unfortunately true that many scientists lack a cultural background, and because of this do not like to look backward. It is a vicious circle: why should they look that way if there is nothing for them to see? Their history of science does not even go as far back as the seventeenth century; they are prone to believe that almost everything worthwhile was done in the nineteenth or in the twentieth century. Now in this they are most certainly wrong. The most astounding results were obtained in the most recent times, simply because they were the latest; but these results were made possible only by all antecedent efforts; they would have been utterly impossible without them. All the preparatory work left undone by our ancestors would have to be done by us now or by our children later. The results of the present are more complex, and more valuable than those of the past, in fact they have superseded the latter; but there is every reason to suppose that in their turn they will be superseded by those of the future. At all times there have been "moderns" who could not help thinking that their ways as compared with those of the "ancients" were almost final. One of the main functions of the history of science is to correct such mistakes and to give us, who are the "moderns" of to-day, a less conceited view of our share in the total of human evolution. Of course this age of ours is a very wonderful one, and for us who are living in it, is undoubtedly for that very reason the most wonderful of all, but we must bear in mind that such privileged ages have succeeded one another as the generations themselves. Even as young lovers have sincerely felt in their exaltation that the world was never more beautiful than as they saw it, even so each great discovery which enabled scientists to penetrate somewhat deeper below appearances and to push the barriers of ignorance and darkness a little further away, may have given them the illusion that they had finally reached the heart of the mystery and that they were the first to understand the universe thoroughly.

There is also a very good practical and philosophical motive for devoting at least as much attention to the more distant achievements as to the later ones, and that is, that the former, although so much easier to explain, give us a far better conception of the meaning of scientific evolution. To begin with they are spread over a much longer period. Modern science, as defined above, is after all hardly more than three centuries old, while the previous evolution was a matter of more than four millennia, that is, without counting the innumerable centuries of which we have no definite records. The development of ancient and mediæval science is not only a much longer stretch, but if I may put it so, a collection of stretches of various lengths interrupted and bent by all kinds of vicissitudes. When we consider the whole of it, we can verify the fact that human evolution is infinitely more complex than the very orderly process of the last centuries would indicate. Scientific research is now organized with such elaboration and in so many countries that a long and complete interruption of it is hardly conceivable, and that we almost expect discoveries to follow each other without cease and without end. In the distant past, on the contrary, there was so much discontinuity and hesitation in scientific progress, that the latter seemed to be even more fortuitous than it really was. A discovery was like a gold nugget that one might stumble upon or not according to one's luck. By way of contrast much of the scientific work of to-day might be compared to the systematic exploitation of a gold mine, the average output of which can be foretold.

That comparison is a little exaggerated on both sides, but the fact remains that scientific progress was far more erratic in the past than it is now, and that considerably more energy was wasted in vain efforts and along hopeless paths. As a result a vision of mediæval man groping for the truth is somewhat bewildering: he seems to be going in too many directions at once and to be turning in circles. There is a general direction however, but to perceive it one must look from a great distance and be able to disregard all the irrelevant movements, all the stops, lapses, detours and retrogressions. We are now sufficiently distant from ancient or even mediæval science to appreciate the meaning of almost every step of it, true or false. On the contrary we cannot yet see the latest developments of science in their true perspective. Of course, we believe we can; we think in good faith that we can single out the most pregnant discoveries of our own days, but the whole of past history is there to testify that contemporary judgments are always precarious. This is natural enough. The value of a theory, the importance of a fact, depend entirely on the conclusions which may be derived from them, on the fruits they will bear, and scientists are not prophets. Comte's saying "Savoir afin de prévoir" is often misquoted. It is true the scientist is able to foresee and to anticipate the immediate consequences of certain events, and therein lies the secret of his material power. But he is not able to predict the future except within the very narrow sector controlled by his knowledge and even there he is hedged in with all kinds of restrictions. Indeed no man is more chary of predictions than the true scientist.

There are two main reasons for studying the history of science: a purely historical one, to analyze the development of civilization, i.e., to understand man, and a philosophical one, to understand the deeper meaning of science. Now from either point of view, the history of ancient and mediæval science is at least as useful as that of modern science. He who knows only one of these histories does not really know the history of science, nor does he know the history of civilization.

I shall try to make this more concrete by dealing at greater length with the earlier parts of our history. If it were not so futile to pick out a single period as the best — for each period was the best from a certain point of view and each was an indispensable link in the chain of ages — I would say in opposition to the uncritical scientist that the most important was, not the latest, but the earliest. Nothing is more difficult than to begin. And what can be more fundamental than a good beginning? Is it not the foundation upon which all the rest will be built?

Unfortunately we shall never have any adequate information on this, the most critical period of man's history, when he was gratifying his urgent needs and slowly emerging out of the darkness, when his instinctive craving for power and for knowledge was beginning to appear. Who first thought of kindling a fire? Who invented the earliest stone implements? Who domesticated the animals which have shared our lives ever since? How did language develop? And later, much later, writing? Who conceived the wheel? Just think of these discoveries and of their infinite implications. Without articulate language man remained an animal. Without writing, the safe transmission and preservation of knowledge were impossible. Progress implies safe keeping of what we already have. Without writing, the accumulation of knowledge was precarious and limited, progress small and uncertain. Can any one of our modern discoveries, however startling, begin to compare with those which made possible all the others? And yet we know nothing about them. We can hardly guess. It is probable that they involved the secular collaboration of thousands of men, each big step forward being finally secured by the exceptional genius of some of them. The evolutions leading to each of these fundamental discoveries were exceedingly slow — almost comparable to the biologic transitions from one type to another — so slow that the people who took part in them were utterly unaware of them. Genius was then required only from time to time to clinch the results obtained by the unconscious accumulation of infinitesimal efforts, to secure what was gained and prepare another slow movement in the same general direction.

The total evolution which prepared the dawn of science must have taken tens of thousands of years. By the beginning of the third millennium before Christ it was already completed in at least two countries: Mesopotamia and Egypt, and possibly in two others, India and China. The people of Mesopotamia and Egypt had then already attained a high stage of culture including the use of writing, and a fair amount of mathematical, astronomical and medical knowledge. Thus it would seem proved that civilization began in the East. *Ex oriente lux, ex occidente lex.* From the East came the light, from the West, law! This aphorism contains a good deal of truth and might be chosen as the motto of my lecture.

Let me say right away that my aim is to show the immense contributions which Eastern people made to our civilization, even if our idea of civilization is focused upon science. We are used to think of our civilization as western, we continually oppose our western ways to the eastern ways, and we have sometimes the impression that the opposition is irreducible.

"Oh, East is East and West is West, and never the twain shall meet."

Now that impression is false, and as it is likely to do considerable mischief in both East and West, it is worthwhile to disclose the error as fully as possible. However divided it may be with regard to material interests and other trifles, mankind is essentially united with regard to its main purpose. East and West are often opposed one to the other, but not necessarily so, and it is wiser to consider them as two visages, or let us say, as two moods of the same man.

Ex oriente lux! There is no doubt whatever that our earliest scientific knowledge is of oriental origin.

As to the possible Chinese and Hindu origins we cannot say much that is definite, but on the contrary with regard to Mesopotamia and Egypt we are on very solid ground.

For example, as early as the middle of the fourth millennium the Egyptians were already acquainted with a decimal system of numbers. In an inscription of that time there is reference to 120,000 captives, 400,000 oxen and 1,422,000 goats, each decimal unit being represented by a special symbol. By the middle of the following millennium Sumerians had developed a highly technical system of accounting. The astronomical knowledge of these people was equally remarkable. The Egyptian calendar of 365 days was established in 4241. Babylonians accumulated planetary observations for astrological purposes; e.g., elaborate observations of Venus go back to the twentieth century. They compiled lists of stars and were soon able to predict eclipses.

That early knowledge was not only abundant, but highly systematized. In the case of Egypt we are especially well informed because we have two early papyri, each of which might be called a treatise. The earliest, the Golenishchev papyrus of Moscow, dates from the middle of the nineteenth century but is copied from an older document of the end of the third millennium; the second, the Rhind papyrus, kept in London and New York, dates from the middle of the seventeenth century but is a copy of a text which may be at least two centuries older. The second of these texts has been studied with extreme care by a number of investigators. The latest edition of it by Arnold Buffum Chace, chancellor of Brown University, Ludlow Bull, H. P. Manning, and R. C. Archibald (1927–29) is at once so complete and so attractive that I am sure it will turn the hearts of many men and women to the study of Egyptian antiquities. I imagine that the first reaction of some people, if they were shown these sumptuous volumes, would be one of wonder that so much time and money should have been spent on an early text of so little scientific value from the point of view of our present knowledge, but I am sure that it would not take long to convert them to an entirely different attitude. For just think what it means. Here we have a mathematical treatise which was written more than thirteen centuries before the time of Euclid! To be sure it does not compare with the latter's Elements, and we are not surprised that more than a millennium of additional efforts were needed to build up the latter, but it contains already such elaborate results that we must consider it, not as a beginning, but rather as a climax, the climax of a very long evolution. The Egyptian mathematicians of the seventeenth century were already able to solve complicated problems involving determinate and indeterminate equations of the first degree and even of the second, their arithmetical ingenuity was astounding, they used the method of false position and the rule of three, they could find the area of a circle and of a sphere with a very remarkable approximation, they could measure the volume of a cylinder and of the frustum of a square pyramid. But is it necessary to insist upon their mathematical accomplishments? Pyramids? Did I not mention pyramids? Do not these gigantic witnesses of the Egyptian genius speak loud enough? The great pyramid of Gizeh dates from the beginning of the thirtieth century. In our age of mechanical wonders, its mass is still as imposing as when it was built almost five thousand years ago; it seems as permanent as the hills and in all probability will outlast most of the skyscrapers of which we are so proud. However startling our first vision of it, our admiration increases as we analyze the achievement and measure the amount of mathematical and engineering skill, of experience and discipline which were needed to bring it to a successful conclusion. No wonder that so many scholars lost their wits for pondering too much on the subject!

If we pass to medicine, other surprises are in store for us. The Greek god of healing, Asclepius, was but a descendant of the Egyptian one, Imhotep, and the history of the latter can be traced back to a real personality, that of a learned physician who flourished probably at the beginning of the thirtieth century. What does this mean again? We often speak of Hippocrates, and we like to call him the Father of Medicine; we shall better appreciate Imhotep's antiquity when we realize that Hippocrates is more than half way between him and us. The chances are that Imhotep's medical knowledge was but rudimentary, but it cannot have been insignificant otherwise his apotheosis would hardly have occurred. However this was only a beginning, or more correctly, a new beginning. Let some thirteen centuries elapse, and we reach the golden age of Egyptian science — the age to which the Rhind papyrus belongs. Strangely enough we have also a medical treatise of the same age, the Edwin Smith papyrus, of which Professor Breasted is preparing an edition. This is not like other papyri, a collection of recipes and charms, but a systematic treatise arranged "*a capite ad calcem*"—from head to foot — an order which was followed down to the end of our Middle Ages. It contains the consideration of forty-eight cases, each of which is reported in the same order: name, examination, diagnosis, judgment, treatment, gloss. We await its complete publication with impatience, but what we already know of it is sufficient to give us a very high idea of early Egyptian medicine and surgery.[1]

These examples will convince you that a considerable body of systematized knowledge was far anterior to Greek science. In fact this helps to explain what one might call the miracle of Greek civilization. To be sure no intelligent man could read the Iliad and the Odyssey, which were the primices of that civilization, without wondering what had made such master-

[1] J. H. Breasted's monumental edition of the Edwin Smith surgical papyrus was finally published in August 1930 (2 vols., University of Chicago Press); it has more than fulfilled our expectations. An elaborate analysis of it appeared in *Isis* (15, 355–67).

pieces possible. They could not possibly appear like bolts from the blue. Like every glorious beginning, this was not only the prelude of one evolution but the end, the climax, of another. Students of Greek mathematics, of Greek astronomy, and Greek medicine could not help asking themselves similar questions. How could the relative perfection of the Greek scientific treatises be accounted for? The explanation is still very incomplete, but no doubt exists as to the main fact: the Greeks borrowed a large quantity of observations and of crude theories from the Egyptians and the peoples of Mesopotamia. Unfortunately it is hardly possible in any case to describe the complete transmission of elements from, say, Egypt to Hellas. This is partly due to the revolutionary events which occurred about the beginning of the first millennium; these events were probably connected with the early use of iron (instead of bronze) and almost obliterated the older Ægean culture. Our ignorance may be dissipated by later archæological discoveries, for example by the deciphering of Minoan and Mycenæan texts, but it is doubtful whether the whole story will ever be revealed to us, for the introduction of the iron age was an upheaval of extraordinary magnitude and destructiveness. At any rate in the present state of our knowledge, there is a gap of more than a thousand years between the golden age of Egyptian science and the golden age of Greek science. We are certain that much of the Greek knowledge was borrowed from eastern sources but we do not know exactly when or how the borrowings took place.

For example, the incubation rites which were practised in the Greek Asclepieia were in all probability derived from Egyptian models. These rites were very important from our point of view because thanks to them a large number of clinical observations were concentrated in the temples, especially in the most famous ones, Epidauros and Pergamon, Cos and Cnidos. The value of such concentration requires no emphasis, least of all for the medical art, for to make scientific inductions, it is not enough to have observations, one must have plenty of them. Without some means of collecting abundant clinical cases as were afforded by the Asclepieia, the progress of medicine would have been considerably slower. It is not too much to say that the Asclepieia were the cradles of Greek medicine, and they help to account for the extraordinary richness of the Hippocratic collection, — but we must not forget that they themselves inherited and continued Egyptian traditions.

On the other hand, Greek astronomy was largely of Babylonian origin, though it was also inspired by Egyptian examples. The Babylonian influence continued to make itself felt throughout historic times, and it is probable that the precession of the equinoxes was first discovered not by Hipparch but by the Babylonian astrologer, Kidinnu (c. 343 B.C.); whether Hipparch borrowed that discovery from Kidinnu or not, it is certain that he could not have made it without reference to the ancient Babylonian observations. With regard to arithmetic, the continuation of Babylonian and Egyptian influences is very striking. The Greek preference for expressing ordinary fractions as the sum of fractions with numerator unity and their use of a special symbol for 2/3 were obviously Egyptian relics, while their sexagesimal fractions were Babylonian.

There is perhaps no more fascinating subject than the study of the transition from oriental science to the early Greek, and the archæological investigations which are being feverishly conducted by scholars of many nationalities all over the Near East are keeping it in a state of flux which is in itself a stimulus. It is perhaps wiser not to indulge in predictions with regard to such a live subject; yet it is safe to say that, however numerous the Greek borrowings may prove to have been, the blossoming of the Greek scientific genius remains almost equally difficult to account for. Students of art and literature are confronted with a similar difficulty, and when we speak of the "Greek miracle" we do nothing but confess to it and admit our ignorance. In fact the difficulty and the miracle are even greater in the case of science than in that of art, for there are Egyptian statues of the early dynasties which are not a whit inferior to the best Greek productions, while the Egyptian scientific treatises, however remarkable especially when their high antiquity is considered, do not begin to compare with their Greek offspring. Between the scribe

Aḥmôse (the writer of the Rhind papyrus) and, say, Hippocrates of Chios, there is such a gigantic difference that some critics have gone so far as to deny the scientific nature of the Egyptian work altogether and to consider it only as a collection of empirical recipes. In this they were certainly mistaken, for the Egyptian knowledge was far from being fragmentary and accidental; it was already methodic to a degree, and hence scientific. Yet the doubts of these critics are somewhat justified by the immensity of the gap. We do not know what happened between the seventeenth and the sixth centuries, and it would be rash to conclude that the Egyptian knowledge was not gradually improved; however the chances are that the main improvements were made not by Egyptians, nor by Minoans or Mycenæans (whoever these were), but by Greeks, the favored people whose earliest "Book" and witness was the Iliad. And these improvements were of such magnitude that they raised science up to a higher level. When a student of ancient science grows a little rhapsodical about it, we may be tempted to ascribe his enthusiasm to the onesidedness and the consequent blindness of his devotion. But I have devoted far more time and thought to the science of the Middle Ages than to that of Antiquity, and my admiration for the latter has not ceased to increase as I knew the former better.

The spirit of Greek science, which accomplished such wonders within a period of about five centuries, was essentially the western spirit, whose triumphs are the boast of modern scientists. But we must bear in mind two important qualifications. First that the foundations of that Greek science were wholly oriental, and however deep the Greek genius it is not certain that it could have built anything comparable to its actual achievements without these foundations. When discussing the fate of a man of genius we may make many suppositions, but it would be absurd to wonder what would have happened if he had other parents, for then he would never have been. In the same way we have no right to disregard the Egyptian father and the Mesopotamian mother of the Greek genius. In the second place, while that genius was creating what might be called (in opposition to Egyptian science on one hand and to mediæval science on the other) the beginning of modern science, another development, equally miraculous, but of an entirely different kind, was taking place in an oriental country near the easternmost end of the Mediterranean Sea. While Greek philosophers were trying to give a rational explanation of the world and boldly postulated its physical unity, the Hebrew prophets were establishing the moral unity of mankind upon the notion of a single God. These two developments were not parallel but complementary; they were equally momentous but entirely independent; in spite of their spatial proximity they proceeded for centuries in almost complete ignorance of one another. They did not really come together until the end of ancient times, and their union was finally cemented upon the prostrate bodies of the two civilizations which had given birth to them.

I shall come back to that presently. But I must first explain the decadence and fall of the Greek spirit. After having made so many conquests in such magnificent style, why did it stop? One cannot help feeling that if that spirit had kept its valor for a few more centuries, human progress would have been considerably accelerated and the course of civilization would have been very different. What befell it? It is impossible to answer such a question, one can only guess, and even our guesses are necessarily very timid. What would we answer in the case of a single man if his best work was done when he was twenty, and the rest of his life spent in sterile idleness? We would say simply: His genius failed him. That would not be a complete explanation but it would satisfy us. But can such explanation hold for a whole nation? Why not? If we speak of the Greek genius at all, as a sort of natural integration, we may conceive the possibility of its gradual corruption and disappearance. If it could emerge, why could it not be submerged again and fail altogether?

What happened to Greece is that the intellectual activities of its people were hopelessly out of proportion to their political wisdom and their morality. A house divided against itself must necessarily fall, a body rent by internal strife is foredoomed to destruction, above all such a body is soon incapable of any kind of creation.[1] It was not simply Greek science that disappeared, but Greek art and literature as well. One might speculate as to what would have happened if the Greek and Hebrew ideals had been nursed together instead of separately, or at any rate, if they had not grown for so long in complete isolation. Such speculations are vain of course, and yet they force themselves upon us. The fact is the Greek and the Hebrew spirits were incompatible; they could not have grown together and corrected one another; rather they would have destroyed each other. After all, it was perhaps necessary that each be built as solidly as possible on its own basis. It is likely that any premature synthesis would have stunted the development of both. When studying the records of the past, one has often the impression that men can grasp but one idea at a time.

The reader knows how Greece was finally conquered by Rome, and how in the course of time it conquered its conquerors. Yet the old spirit was subdued, and Roman science even at its best was always but a pale imitation of the Greek. The Romans were so afraid of disinterested research, the excess of which had been one of the causes of the Greek corruption, that they went to the other extreme and discouraged

[1] The Euripidean quotation on p. 38 is typical, for it betrays political indifference as well as scientific interest. The Greeks carried their political sluggishness and immorality so far that they ceased to exist as a nation, and jeopardized not only their political but also their intellectual life.

any research, the utilitarian value of which was not immediately obvious.

In the meanwhile Jesus Christ had appeared and told the world a new message, a message of love and humility, universal in its scope: Charity does not need knowledge; blessed are the pure in spirit, the pure in heart; on the other hand, knowledge without charity is not only useless but pernicious; it can but lead to pride and damnation. The development of Christianity was a first attempt to bring together the Hebrew and the Greek spirits, but as the Roman Christians hardly understood the former and misunderstood the latter thoroughly, the attempt was an utter failure.

A good example of those misunderstandings may be found in the work of Tatian, a Syrian convert who lived in Galen's time. His Greek oration "against the Greeks" contains not only an account of the weaknesses of paganism but the most extravagant claims in behalf of oriental peoples. According to him the Greeks had invented nothing; they had borrowed all their knowledge from others — Assyrians, Phœnicians, Egyptians; their only superiority was in the art of writing and of lying. Thus after centuries of ignorance of Eastern achievements, some Eastern Greeks, whose minds were poisoned against Greek civilization by Christian prejudices, were going to the other extreme. Apparently Greeks and Orientals were not fated to understand one another.

We may say that the Greek spirit, that disinterested love of truth which is the very spring of knowledge, was finally smothered by the combination of Roman utilitarianism and Christian sentimentality. Again let us dream for a moment, and wonder what might have happened if the Greeks and the Christians had seen their respective good points instead of seeing only the evil ones. How beautiful if their two types of otherworldliness could have been harmonized! How many miseries mankind would have been spared! But it was not to be. The path of progress is not straight but very crooked; the general direction is clear enough but only if one considers a very long stretch of it from far off. Before being able to reconcile the love of truth with the love of man, the scientific spirit with the Golden Rule, mankind was obliged to make many strange and cruel experiments.

To begin with, under the influence of Christian education combined with Roman narrowmindedness and later with Barbarian ignorance, the connection with the Greek culture — which was the only source of positive knowledge — became looser and looser. The debasement of thought is well illustrated by the fact that even in the Byzantine empire, where there was no linguistic barrier to the transmission of ancient science, much of the latter remained practically unknown. This is so true that in the thirteenth and fourteenth centuries, when the Latin world was finally awakened, Byzantine scholars preparing a scientific revival retranslated from the Arabic and the Latin a number of writings which were nothing but translations from the Greek or poor imitations of such translations. Their intellectual indigence was such that they did not recognize the work of their own ancestors.

The contact between ancient Greece and western Christendom ended by being so precarious that it might have conceivably been broken altogether, but for the intervention of another oriental people, the Arabs. Please note that this was the third great wave of oriental wisdom, the third time that the creative impulse came from the East. The first initiative — and the most fundamental of all — came from Egypt and Mesopotamia; the second from Israel, and though it influenced science only in an indirect way, it was also of incalculable pregnancy; the third, with which I am going to deal now, came from Arabia and from Persia.

About the year 610 a new prophet appeared at Mecca in Hejaz, Abû-l-Qâsim Muhammad of the tribe of Quraysh, who was like a new incarnation of the old Hebrew prophets. At first the people did not pay much attention to him, but after he had abandoned his native town and moved two hundred and fifty-five miles northward to al-Medina, in 622, his success was phenomenal. No prophet was ever more successful. By the time of his death ten years later he had managed to unite the Arabian tribes and to inspire them with a single-hearted fervor which would enable them later to conquer the world. Damascus was captured in 635, Jerusalem in 637; the conquest of Egypt was completed in 641, that of Persia in the following year, that of Spain somewhat later in 710/12. By this time the Muslims, that is, the Prophet's followers, were ruling a large belt of the world all the way from Central Asia to the Far West. The conquest of Persia was especially momentous because it brought the invaders, brave but uncouth, into touch with an old and very refined civilization, that of Iran. I did not speak of it before because it is difficult to state its earlier contributions with sufficient brevity and more difficult, if not impossible, to date them. For the purpose of a sketch like this, it is sufficient to introduce Iran at this juncture, but its part henceforth was considerable. The new dynasty of Muslim caliphs, the 'Abbāsid (750–1258) established its capital in Baghdād on the Tigris, and for a time that new city was one of the main centers of the civilized world. The 'Abbāsids were from the beginning under the Iranian spell. Their religious and moral strength was derived from their ancestral home, Arabia; their urbanity, their humanism, from Persia. To put it in a nutshell the new Muslim civilization was essentially due to the grafting of the vigorous Arabic scion upon the old Iranian tree. This explains at once its astounding robustness and its changing qualities.

Under the impulse of these two tremendous forces, Muslim fanaticism and Persian curiosity, and under the guidance of a series of 'Abbāsid caliphs who had a passion for knowledge — al-Manṣūr, Hārūn al-

Rashīd, al-Ma'mūn — the new civilization developed with incredible speed and efficacy. It was doubly rooted in the past: The Prophet had transmitted to them with very few modifications Semitic monotheism and morality, and their Persian tutors had incited them to drink deeply into the older sources of learning, Sanskrit and Greek. From the Hindus they learned arithmetic, algebra, trigonometry, iatrochemistry; from the Greeks, logic, geometry, astronomy, and medicine. It did not take them long to realize the immensity of the Greek treasure and they had no rest until the whole of it (that is, as much as was available to them) was translated into Arabic.

In this enterprise they received invaluable help from the Syrians and other Christian subjects of the Caliphate who spoke Greek, Syriac, and pretty soon Arabic. These oriental Christians, though somewhat Hellenized, had always been treated with suspicion and disfavor by the Byzantine government, and if (as is very probable) they shared Tatian's views, it is not surprising that no love was lost between them. Being repulsed and persecuted by the Greeks, their readiness to help their Muslim conquerors was not astonishing. The Syrians spoke Arabic with so much alacrity that they gradually allowed this new language to supersede their own. These born polyglots were natural intermediaries; it is they who prepared the earliest translations from the Greek into Arabic and who initiated their masters in the Greek knowledge. Thus were the first bridges between Hellas and Islām built by Christians.

The immense cultural importance of Islām lies in the fact that it brought finally together the two great intellectual streams which had flowed independently in ancient times. Previous attempts, as I have already indicated, had failed. Jews and Greeks had mixed in Alexandria but in spite of the fact that the former had learned the language of the latter and that one of their learned men, Philon, had made a deep study of both traditions, there had been no real fusion. The Christians had not succeeded any better, because of their single-hearted devotion to the new Gospel, which reduced everything else to futility in their eyes. Now for the first time in the history of the world Semitic religion and Greek knowledge actually combined in the minds of many people. Nor was that integration restricted to a single city or country; the new culture spread like a prairie fire from Baghdād eastward to India, Transoxiana and further still, and westward to the very edge of the world.

Muslim culture was at once deeply unified and very diversified. The peoples of Islām were kept together and separated from the rest of the world by the two strongest bonds which can bind a human community, religion and language. One of the few duties of a learned Muslim is the reading of the Qur'ān (their Bible), and it must be read in Arabic. Thanks to this religious obligation, Arabic, which before Muhammad had no more than a tribal importance, became a world language. After the eleventh century it lost its hegemony, but remained very important; it is still one of the languages most widely used at the present time. It has gradually been split into a number of dialectal forms, even as Latin disintegrated itself into the various Romance languages, but with these radical distinctions that up to this day, every literate Muslim must have some knowledge of classical Arabic to read the Qur'ān, and that the written language — e.g., that used in newspapers — continues to approximate more or less the classical standards. While each Romance language has its own written form, its own standards of perfection, one may say that there is for the Arabic writer all over the world but one model of excellence, that given by the Qur'ān and by the best authors of the classical age. Because of their single language and of their common faith,[1] ideas traveled with astounding regularity and speed from one end of the Dār al-Islām to the other.

The universal extension of that culture caused necessarily many diversities. Muslims were brought closely into touch with all kinds of unbelievers, — in

[1] To be sure, Islām was soon divided into a number of sects and schools, and one finds in it the same gamut of religious forms as in Christianity, — from the extreme fundamentalism and the strangest mystical aberrations at the right to the purest unitarianism at the left; yet, however different, these were all forms of the same Muslim faith. Every Muslim read the same Scriptures.

the East, Chinese, Mongols, Malays, Hindus; further west, Magians, Syrians, Greeks, Copts, further still, Berbers in Africa; Sicilians, Spaniards, and other Franks in southern Europe; Jews, everywhere. These contacts were generally friendly, or at least not unfriendly, for the Muslims treated their ra'āyā (subjects) with kind and tolerant condescension. Under their patronage, many important works were published in Arabic by non-Muslims: Sabians, Christians, Jews, Samaritans. The great chemist, Jābir ibn Ḥaiyān, was probably a Sabian; al-Battānī was certainly of Sabian origin but had embraced Islām; the physicians Ḥunain ibn Isḥāq, Ibn Buṭlān and Ibn Jazla were Christians. Down to the twelfth century Arabic was the philosophic and scientific language of the Jews; for example, the famous Guide of the Perplexed, the greatest Jewish treatise of the Middle Ages, was written by Maimonides in Arabic. What is more, the earliest Hebrew grammars were composed also in Arabic, not in Hebrew. In other words the mediæval Jews were so deeply Arabicized, that they needed Arabic assistance for the scientific study of their own sacred language.[1]

During the first two centuries of the Hegira the whole of Islām was ruled by the Ummayad and 'Abbāsid caliphs, but after that the caliphate was gradually broken into an increasing number of independent

[1] In a similar way, American Jews study Hebrew grammar in English books, but the analogy ends there. Hebrew grammar was actually born in an Arabic cradle. See my *Introduction to the History of Science* (vol. 1, 1927, pp. 623, 653, and by index *sub voce* Hebrew grammar).

kingdoms of all kinds and sizes. The political disintegration caused intense rivalries, intellectual ones as well as others, between the different Muslim courts. Instead of one or two centers of culture, like Baghdād and Cordova, there grew up little by little a whole series of them: Ghazna, Samarqand, Marv, Herât, Ṭūs, Nīshāpūr, Ray, Isfahān, Shīrāz, Mūṣul, Damascus, Jerusalem, Cairo, Qairawān, Fās, Marrākush, Toledo, Seville, Granada, etc., etc. The obligation for every Muslim to perform if possible the Pilgrimage to Mecca brought about incessant communications between the different parts of Islām and originated numberless personal meetings between scholars hailing from the most distant countries. Under that influence many learned Muslims seemed to be affected with a kind of *Wanderlust*, for it was not unusual for them to perform the Pilgrimage more than once, making considerable stops in the main cities on their way, renewing contacts with their colleagues, engaging in long discussions, copying manuscripts, or composing their own writings, this one in Andalusia, another in the Maghrib, another in Egypt, and so forth. Thus (and also because of the common language) scientific knowledge obtained in any part of Islām was transmitted with astounding celerity to the others, and fresh stimulations were constantly exchanged.

The almost unbelievable vigor of the new culture may be well measured by the international triumph of the Arabic language, a triumph which was the more remarkable because that language was not ready for the occasion but had to be elaborated as the need for it increased and became more and more technical. The Qur'ānic idiom was very beautiful indeed but limited. As the immense task of pouring out the Greek treasure into the Arabic vessels proceeded, it was necessary to make new vessels and better ones. Not only that but a great majority of the people who used them had to begin by learning how from the very rudiments. And yet within a couple of centuries multitudes had acquired some familiarity with that language which had been utterly unknown to their ancestors, if not to their own parents.

The briefest enumeration of the Arabic contributions to knowledge would be too long to be inserted here, but I must insist on the fact that, though a major part of the activity of Arabic-writing scholars consisted in the translation of Greek works and their assimilation, they did far more than that. They did not simply transmit ancient knowledge, they created a new one. To be sure none of them attained unto the highest peaks of the Greek genius. No Arabic mathematician can begin to compare with Archimedes or Apollonius. Ibn Sīnā makes one think of Galen, but no Arabic physician had the wisdom of Hippocrates. However, such comparisons are hardly fair, for a few Greeks had reached almost suddenly extraordinary heights. That is what we call the Greek miracle. But one might speak also, though in a different sense, of an Arabic miracle. The creation of a new civilization of international and encyclopædic magnitude within less than two centuries is something that we can describe, but not completely explain. This movement, as opposed to the Greek, was perhaps more remarkable for its quantity than for its quality. Yet it was creative; it was the most creative movement of the Middle Ages down to the thirteenth century. The Arabic-writing scientists elaborated algebra (the name is a telltale) and trigonometry on Greco-Hindu foundations; they reconstructed and developed — though, it must be said, very little — Greek geometry; they collected abundant astronomical observations and their criticisms of the Ptolemaic system, though not always justified, helped to prepare the astronomical reformation of the sixteenth century; they enriched enormously our medical experience; they were the distant originators of modern chemistry; they improved the knowledge of optics, and meteorology, the measurement of densities; their geographical investigations extended from one end of the world to the other; they published a number of annals of capital interest, dealing with almost every civilized country outside of western Christendom; one of their historians, the Berber Ibn Khaldūn, expounded a philosophy of history which was by far the most elaborate and the most original of mediæval times; finally they laid down the principles of Semitic philology.

Surely these were no mean achievements. If they lacked the supreme quality of the best ancient efforts, we must remember that few men have ever come as near to perfection as the best of the Greeks. On the other hand, if we place them in their own environment and compare the Arabic with other mediæval efforts, the immense superiority of the former is obvious. We may say that from the middle of the eighth century to the end of the eleventh, the Arabic-speaking peoples (including within their ranks, it is true, a number of Jews and Christians) were marching at the head of mankind. Thanks to them Arabic was become not only the sacred language of the Qur'ān, the vehicle of God's own thoughts, but the international language of science, the vehicle of human progress. Just as to-day the shortest way to knowledge for any Oriental, is the mastery of one of the main occidental languages, even so during these four centuries Arabic was the key, and almost the only key, to the new expanding culture.

Indeed the superiority of Muslim culture, say in the eleventh century, was so great that we can understand their intellectual pride. It is easy to imagine their doctors speaking of the western barbarians almost in the same spirit as ours do of the "Orientals." If there had been some ferocious eugenists among the Muslims they might have suggested some means of breeding out all the western Christians and the Greeks because of their hopeless backwardness. At that time Muslim pride would have been the more conceivable because they had almost reached their climax, and pride is never as great as when the fall is near. On the contrary only a few Christians were then aware of their inferiority; that awareness did not come upon them until much later — by the middle of the thirteenth century — when Islām was already on the downward path and Latin Christendom was climbing higher and higher. This is very interesting, but the rule rather than the exception: when people boast too much of their culture it means either that it is so new that they have not yet grown accustomed to it or else that it is already decadent and that they try to hide (even from themselves) their incompetence under the cloak of past achievements. In the thirteenth century Islām was in the decadent and boasting stage, while Christendom had finally realized the richness of the Greco-Arabic knowledge and made gigantic efforts to be allowed to share it, and hence was relatively in a chastened mood.

For the sake of illustration let us consider the levels of mathematical knowledge among Muslims and among Christians in the first half of the eleventh century. There was then a splendid mathematical school in Cairo, made famous by the great astronomer Ibn Yūnus and the great physicist Ibn al-Haitham; al-Karkhī was working in Baghdād, Ibn Sīnā in Persia, al-Bīrūnī in Afghānistān. These mathematicians and others, like Ibn al-Ḥusain and Abū-l-Jūd, were not afraid to tackle the most difficult problems of Greek geometry; they solved cubic equations by the intersection of conics, they investigated the regular heptagon and enneagon, developed spherical trigonometry, Diophantine analysis, etc. Pass to the West and what do we find? Wretched little treatises on the calendar, on the use of the abacus, on Roman (duodecimal) fractions, etc. We have a "mathematical" correspondence exchanged (c. 1025) by two schoolmasters, Ragimbold of Cologne and Radolf of Liége. It is truly pitiful. Their geometry was on the pre-Pythagorean level; they were not bad computers, it is true; we might compare them to the Egyptian scribe Aḥmôse, who had done his task almost twenty-seven centuries before!

How is it that the Muslim or oriental supremacy ended about the end of the eleventh century? There was a double cause to this: the Arabic genius was less vigorous and less fertile; the power and knowledge of the Latin world was growing faster and faster. The Arabic achievements did not stop, not by any means. Great Arabic scientists and scholars continued to appear until the fourteenth century and even later. For example, mathematicians and astronomers like Jābir ibn Aflaḥ, al-Biṭrūjī, al-Ḥasan al-Marrākushī, Nāṣir al-Dīn al-Ṭūsī; physicists like al-Khāzinī, Quṭb al-Dīn al-Shīrāzī, Kamāl al-Dīn ibn Yūnus; geographers like Yāqūt, al-Qazwīnī, Abū-l-Fidā', Ibn Baṭṭūṭa; philosophers like Ibn Rushd, Fakhr al-Dīn al-Rāzī, 'Abd al-Laṭīf; physicians like Ibn Zuhr and Ibn al-Baiṭār; botanists and agriculturists like Ibn al-Ṣūrī and Ibn al-'Awwām; historians like Ibn Khallikān, Rashīd al-Dīn, Ibn Khaldūn, al-Maqrīzī, etc., etc. This list might be lengthened considerably and yet contain only very distinguished names; as it is, it includes some of the most illustrious ones in the whole history of civilization. The men I have mentioned hailed from every part of Islām; a few of them wrote in Persian, but even for those Arabic was a privileged language. Yet by the end of the eleventh century the main task of the Arabic scientists — as far as it concerned the whole world and not only themselves — was already completed, and after that time the relative importance of Muslim culture declined steadily. During the twelfth century its prestige was due even more to its past than to its present achievements, great as these were. In the meanwhile Christians and Jews were feverishly pouring out the Greco-Arabic learning from the Arabic vessels into the Latin and Hebrew ones.

The Christians were far ahead of the Jews in this new stage of transmission, and that for a very simple reason. Down to the eleventh century the philosophic and scientific (as opposed to the purely rabbinical) activities of the Jews were almost exclusively confined to the Muslim world. The Jewish philosophers, grammarians, scientists who lived under the protection of Islām were generally well treated, and some of them — like Ḥasdai ibn Shapruṭ in Cordova — attained positions of high authority and became the intellectual as well as the political leaders of their time. These Jews of the Dār al-Islām were bilingual; Hebrew was of course their religious language and probably also their domestic one, but for all philosophic and scientific purposes they thought in Arabic. They had no need of translations. On the contrary it was much easier for them to read a medical book in Arabic than in Hebrew. Sometimes they would copy Arabic manuscripts in Hebrew script, but even that was not really indispensable; it was more a matter of convenience than of necessity.

On the other hand as soon as the Latin Christians began to realize the importance of the Arabic literature, as only a few of them could ever hope to master a language as alien to their own and written in such illegible and mystifying script, they longed for translations and did all they could to obtain them. By the end of the eleventh century their longing was partly fulfilled by Constantine the African, aptly called "magister orientis et occidentis"; he was indeed one of the great intermediaries between the East and the West. Constantine translated a large number of Greco-Muslim works from Arabic into Latin at the monastery of Monte Cassino, where he died in 1087. As we might expect, the results of his activity, far from appeasing the hunger of European scholars, stimulated it considerably. It now dawned upon the most advanced of them that the Arabic writings were not simply important but essential, for they contained vast treasures of knowledge, the accumulated learning and experience of the whole past. It is no exaggeration to say that during the twelfth century and down to about the middle of the thirteenth century the foremost activity of Christian scholars was the translation of Arabic treatises into Latin. There appeared a succession of translators of such size that they have almost the dignity of creators: Adelard of Bath, John of Seville, Domingo Gundisalvo, and many others, including the greatest of all times, Gerard of Cremona. By the end of the twelfth century, the main body of Greco-Arabic knowledge was already available to Latin readers, but the more they had, the more they wanted. By the end of the following century, and even by the middle of it, there was little of real importance in the Arabic scientific literature which they were not aware of. Moreover under the stimulus of the Arabic writings, some translators took pains to rediscover the Greek originals, and their translations straight from the Greek followed closely upon the heels of those from the Arabic. A remarkable case is that of the *Almagest*. This was actually translated from the Greek before being translated from the Arabic; the direct translation was made in Sicily c. 1160, the indirect one was completed by Gerard of Cremona at Toledo in 1175. Yet such was the strength of the Arabic tradition and Gerard's own prestige, that the earlier version though presumably better was entirely superseded by the second.

At first the eastern Jews and those of Spain were much better off than the Christians for the whole of Arabic literature was open to them without effort, but in the twelfth century the scientific life of Judaism began to move from Spain across the Pyrenees, and in the following century it began to decline in its former haunts. By the middle of the thirteenth century a great many Jews had already been established so long in France, Germany, England, that Arabic had become a foreign language to them. Up to this period the Jews had been generally ahead of the Christians, and far ahead; now for the first time the situation was reversed. Indeed the Christians had already transferred most of the Arabic knowledge into Latin; the translations from Arabic into Hebrew were naturally far less abundant, and hence the non-Arabic speaking Jews of western Europe were not only in a position of political inferiority (the Crusades had caused many anti-Semitic persecutions and the Jews of Christendom were everywhere on the defensive) but also—and this was perhaps even more painful — in a position of intellectual inferiority. To be sure this was soon compensated by the fact that many of them learned Latin and could then read the Arabic texts in their Latin versions, but even then they did not have any longer an intellectual monopoly against the Christians; they came but second. While the early Jewish physicians had possessed "secrets" of learning which were sealed to their Christian colleagues (this was especially true with regard to eye-diseases which were thoroughly investigated in Arabic treatises), the later ones had no such privileges. The gravity of the change is well illustrated by the appearance in the fourteenth and following centuries of an increasing number of translations (e.g., of medical works) from Latin into Hebrew. Thus the stream of translations which had been running from East to West was again reversed in the opposite direction. Note that a curious cycle had been completed, for the source of these writings was Greek; their Arabic elaborations had been translated into Latin and had inspired new Latin treatises; these treatises were now translated into Hebrew. From East to East via the West! But other cycles were even more curious. In the fourteenth century and later, Arabic, Persian, and Latin writings which were ultimately of Greek origin were retranslated into Greek. For example, the most popular logical textbook of the Middle Ages, the *Summulæ logicales* of Peter of Spain [1] was not only translated into Hebrew, but also into the very language from which its main sustenance had been indirectly derived. From Greek to Greek via Arabic and Latin!

Incidentally this will help the reader to realize the usefulness of studying ancient translations. These give us the best means of appreciating the relative levels of various civilizations at definite periods. We can watch their rise and fall and so to say measure them. Streams of knowledge are constantly passing from one civilization into the others, and in the intellectual even as in the material world, streams do not run upward. From a single translation one could deduct nothing, for its occurrence might be erratic. In the past even as now it was not necessarily the

[1] Pope John XXI.

best writings which were translated; indeed some of the worst enjoyed that distinction more than any others. But if we consider the whole mass of translations, we can reconstruct the cultural exchanges and draw conclusions of the greatest interest. To return to my comparison of mankind with a single man, the activity of translators helps us to evoke the intellectual evolution of the former: we can tell which was the dominating influence at each time, and so to say retrace his wandering steps across the schools and the academies of the old world.

During the twelfth century the three civilizations which exerted the deepest influence upon human thought and which had the largest share in the molding of the future, the Jewish, the Christian, and the Muslim, were remarkably well balanced, but that state of equilibrium could not last very long, because it was due to the fact that the Muslims were going down while the two others were going up. By the end of the twelfth century it was already clear (that is, it would have been clear to any outside observer, as it is to ourselves) that the Muslims would soon be out of the race, and that the competition would be restricted to the Christians and the Jews. Now the latter were hopelessly jeopardized by their political servitude and by the jealous intolerance and the utter lack of generosity (to put it mildly) of their rivals. Moreover for the reason explained above the main sources of knowledge were now less available to them than to their persecutors. This went much deeper than it seems, for when an abundant treasure of knowledge becomes suddenly available to a group of people, it is not only the knowledge itself that matters, but the stimulation following in its wake. The Jews were steadily driven into the background, and in proportion as they were more isolated they tended to increase their isolation by devoting their attention more exclusively to their own Talmudic studies.

Toward the end of the thirteenth century some of the greatest doctors of Christendom, Albert the Great, Roger Bacon, Ramon Lull, were ready to acknowledge the many superiorities of Arabic culture. It is paradoxical but not surprising that at the very time when that full realization had come to them, that culture was already declining, and their own was finally triumphing. From that time on, the Christians enjoyed the political and intellectual hegemony. The center of gravity of the learned world was in the West and it has remained there until our own days; by a strange irony of fate it may even pass some day beyond the western ocean which was then supposed to be an insuperable barrier. Moreover because of the decadence and fall of Muslim Spain and of the growing isolation and aloofness of the Jews, the West became more and more westernized. Of course Muslim and Jewish efforts went on and both faiths produced many great men in the following centuries, yet the western supremacy continued to wax until a time was reached, in the sixteenth century, when the expanding civilization was so deeply westernized that the people — even those of the Orient — began to forget its oriental origins, and when the very conception of Muslim and Jewish science almost disappeared. That conception may seem artificial to us, but I believe I have made it clear enough that it was a perfectly natural and necessary one in mediæval times. The final results of science are of course independent of the people who discovered them, but we are anxious to find out how much we owe to each of them, in what kind of environment knowledge developed, and which devious ways the human spirit followed throughout the ages. After the sixteenth century, when science was finally disentangled from theology, the distinctions between Jewish, Christian, and Muslim science ceased to be justified, but it keeps its historical value. In spite of his deep Jewishness and of his abundant use of Jewish sources, we do not count Spinoza any more as a Jewish philosopher in the same sense as we counted Maimonides or Levi ben Gershon; he is one of the founders of modern philosophy, one of the noblest representatives of the human mind, not eastern or western, but the two unified.

Perhaps the main, as well as the least obvious, achievement of the Middle Ages, was the creation of the experimental spirit, or more exactly its slow incubation. This was primarily due to Muslims down to the end of the twelfth century, then to Christians. Thus in this essential respect, East and West cooperated like brothers. However much one may admire Greek science, one must recognize that it was sadly deficient with regard to this (the experimental) point of view which turned out to be the fundamental point of view of modern science. Though their great physicians followed instinctively experimental methods, these methods were never properly appreciated by their philosophers or by the students of nature. A history of the Greek experimental science, outside of medicine, would be exceedingly short. Under the influence of Arabic alchemists and opticians and later of Christian mechanicians and physicists the experimental spirit grew very slowly. For centuries it remained very weak, comparable to a delicate little plant always in danger of being ruthlessly trampled down by dogmatic theologians and conceited philosophers. The tremendous awakening due to the western re-discovery of printing and to the exploration of the new world, accelerated its development. By the beginning of the sixteenth century it was already lifting its head up, and we may consider Leonardo da Vinci its first deliberate vindicator. After that its progress became more and more rapid, and by the beginning of the following century, experimental philosophy was admirably explained by another Tuscan, Galileo, the herald of modern science.

Thus if we take a very broad view of the history of science, we may distinguish in it four main phases. The first is the empirical development of Egyptian and Mesopotamian knowledge. The second is the building of a rational foundation of astounding beauty and strength by the Greeks. The third and until recently the least known, is the mediæval period — many centuries of groping. Immense efforts were spent to solve pseudo-problems, chiefly to conciliate the results of Greek philosophy with religious dogmas of various kinds. Such efforts were naturally sterile, as far as their main object was concerned, but they brought into being many incidental results. The main result, as I have just explained, was the incubation of the experimental spirit. Its final emergence marks the transition between the third period and the fourth, which is the period of modern science. Note that out of these four periods the first is entirely oriental, the third is mostly but not exclusively so; the second and fourth are exclusively western.

To return to the fourth period — which is still continuing — the final establishment of the experimental philosophy was indeed its main distinction, its standard, and its glory. Not only did the new method open the path to untold and unimaginable discoveries, but it put an end to unprofitable quests and idle discussions; it broke the vicious circles wherein philosophers had been obstinately turning for more than a thousand years. It was simple enough in itself, but could not be understood as long as a series of intellectual prejudices obscured man's vision. It may be summed up as follows: Establish the facts by direct, frequent, and careful observations, and check them repeatedly one against the other; these facts will be your premises. When many variables are related find out what happens when only one is allowed to vary, the others remaining constant. Multiply such experiments as much as you can, and make them with the utmost precision in your power. Draw your conclusions and express them in mathematical language if possible. Apply all your mathematical resources to the transformation of the equations; confront the new equations thus obtained with reality. That is, see what they mean, which group of facts they represent. Make new experiments on the basis of these new facts, etc., etc.

All the triumphs of modern science have been due to the application, more or less deliberate, of that method. Moreover experimental scientists have laid more and more emphasis on the needs of objective verification. Truth is relative but it becomes less and less so, and more and more reliable, in proportion as it has been checked oftener and in a greater variety of ways. The experimental method, simple as it may seem to anyone who approaches it with an open mind, developed only very gradually. Little by little scientists learned by experience to trust their reason more than their feelings, but also not to trust their reason too much. The results of any argument, just like the results of any mathematical transformation, are not entirely valid until they have been checked and rechecked in many ways. Facts can only be explained by theories, but they can never be explained away; thus, however insignificant in themselves, they remain supreme. They are like the stones of a building; individual stones are worthless but the building would have no reality without them.

It is amusing to hear the old humanists speak of restraint and discipline as if they had the monopoly of these qualities, when the experimental method is itself the most elaborate discipline of thought which has ever been conceived. To be sure it does not apply to everything; nor does it claim any monopoly for itself except within its own domain. It is the experimental method which has given to human reason its full potency, but at the same time it has clearly shown its limitations and provided means of controlling it. It has proved the relativity of truth, but at the same time has made it possible to measure its objectivity and its degree of approximation. Above all it has taught men to be impartial (or at least to try to be), to want the whole truth, and not only the part of it which may be convenient or agreeable. Such impartiality was obviously impossible, and almost unconceivable, as long as the objectivity of truth could not be appreciated.

The experimental method is in appearance the most revolutionary of all methods. Does it not lead to astounding discoveries and inventions? Does it not change the face of the world so deeply and so often that superficial people think of it as the very spirit of change? And yet it is essentially conservative, for it hesitates to draw conclusions until their validity has been established and verified in many ways; it is so cautious that it often gives an impression of timidity. It seems revolutionary because it is so efficient; its conclusions because of their restraint cannot be opposed; because of their strength they cannot be frustrated. When thought is as severely disciplined as scientific thought is, it is irresistible, and yet it is the greatest element of stability in the world. How shall we account for that paradox? Progress implies stability; it implies the respect of traditions. Scientific thought is or seems revolutionary because the consequences it leads to are so great and often unexpected, but it leads to them in a steady way. The history of science describes an evolution of incomparable magnitude which gives us a very high idea of man's intellectual power, but that evolution is as steady as that which is caused by natural forces. You have heard the story of the cowboy, who coming suddenly upon the rim of the Grand Cañon, exclaimed: "Good Lord, something has happened here!" Now, as you know, the cowboy was wrong if he meant that something had suddenly happened at a definite time, and had been rapidly completed. In that sense nothing ever happened in the Grand Cañon. In the same way the development of science, though incomparably swifter than the cutting of a cañon, is a steady process; it seems revolutionary, because we do not really see the process, but only the gigantic results.

From the point of view of experimental science, especially in its present stage of development, the opposition between East and West seems extreme. However — and this is the burden of my lecture — we must bear in mind two things.

The first is that the seeds of science, including the experimental method and mathematics, in fact the seeds of all the forms of science came from the East, and that during the Middle Ages they were largely developed by Eastern people. Thus in a large sense, experimental science is a child not only of the West, but also of the East; the East was its mother, the West was its father.

In the second place I am fully convinced that the West still needs the East to-day, as much as the East needs the West. As soon as the Eastern peoples have unlearned their scholastic and argumentative methods, as we did in the sixteenth century, as soon as they are truly inspired with the experimental spirit, there is no telling what they may be able to do for us, or, heaven forbid! against us. To be sure, as far as scientific research is concerned they could only work with us, but their applications of it might be very different. We must not make the same mistake as the Greeks who thought for centuries that their spirit was the only one, who ignored altogether the Semitic spirit and considered foreign people barbarians; their ultimate fall was as deep as their triumph had been high. Remember the rhythm between East and West; many times already has our inspiration come from the East; why should that never happen again? The chances are that great ideas will still reach us from the East and we must be ready to welcome them.

The men who assume a truculent attitude against the East and make the most extravagant claims for the Western civilization, are not likely to be scientists. Most of them have neither knowledge nor understanding of science; that is, they do not in the least deserve the superiority of which they boast so much and which their incoherent desires would soon extinguish, if they were left to themselves.

We are justly proud of our American civilization, but its records are still very short. Three centuries! how little that is as compared with the totality of human experience; hardly more than a moment, a wink of the eye. Will it last? Will it improve or wane and die out? There are many unhealthy elements in it and if we wish to uproot them before the disease has spread beyond our control, we must expose them mercilessly, but that is not my task. If we want our civilization to justify itself we must do our best to purify it. One of the best ways of doing this is the cultivation of disinterested science; the love of truth — as a scientist loves it, the whole of it, pleasant or unpleasant, useful or not; the love of truth, not the fear of it; the hatred of superstition, no matter how beautiful its disguises may be. Whether our civilization will last or not, at any rate it has not yet proved its longevity. Hence we must be modest.

After all the main test is that of survival, and we have not yet been tried.

New inspirations may still, and do still, come from the East, and we shall be wiser if we realize it. In spite of its prodigious triumphs, the scientific method is not all-sufficient. It is supreme when it can be applied and when it is well applied, but it would be foolish not to recognize the two kinds of limitations which this implies. First, the method cannot always be applied. There are immense realms of thought where it is thus far inapplicable — art, religion, morality. Perhaps it will always be inapplicable to them. Second, it can be very easily misapplied, and the possibilities of misapplication of such an inexhaustible source of power are appalling. Think of the war when thousands of people used their scientific knowledge and their ingenuity to devise new means of destruction, when the whole machine was working in reverse. Happily the war stopped when recovery was still easy, but our historical experience proves that it could have been carried further, to the verge of annihilation.[1] I believe with Mr. Robert A. Millikan[2] that the development of science is constantly decreasing the probability of war but it does not reduce it to naught, and on the other hand it tends to increase its

[1] Witness the war between Paraguay and her neighbors in 1864–70. At the beginning of that war the population of Paraguay was about 1,337,439; at the end it was reduced to 221,079, that is, 106,254 women above fifteen, 86,079 children and 28,746 men. Figures quoted by G. M. McBride in *Encyclopædia Britannica* (vol. 17, p. 359, 1929).

[2] *Science and the New Civilization* (New York, Scribner's, 1930). See *Isis* 14, 446–49.

destructiveness and its extent. The risk of war may be much smaller, but the cataclysm is likely to be more devastating if it occurs at all. Thus the danger of war and of other perversions of our technical power remains considerable.

It is clear that the scientific spirit is unable to control its own applications. To begin with these applications are often in the hands of people who have no scientific knowledge whatever; for example, it is not necessary to have any education or instruction in order to drive a high-powered car which may cause any amount of destruction. But even scientists might be tempted to misapply their knowledge under the influence of a strong passion. The scientific spirit must be itself assisted by other forces of a different kind, — by religion and morality. In any case, it must not be arrogant, nor aggressive, for it is like all other things human, essentially imperfect.

The unity of mankind includes East and West. They are like two moods of the same man; they represent two fundamental and complementary phases of human experience. Scientific truth is the same East and West, and so are beauty and charity. Man is the same everywhere with a little more emphasis on this or that.

East and West, who said the twain shall never meet? They meet in the soul of every great artist who is more than an artist and whose love is not restricted to beauty; they meet also in the soul of every great scientist who has been brought to realize that truth, however precious, is not the whole of life, that it must be completed by beauty and charity.

Let us remember with gratitude all that we owe to the East — the moral earnestness of Judæa, the Golden Rule, the very rudiments of the science we are so proud of — this is an immense debt. There is no reason why it should not be indefinitely increased in the future. We must not be too sure of ourselves; our science may be great, our ignorance is greater still. By all means let us develop our methods, improve our intellectual discipline, continue our scientific work, slowly, steadily, in a humble spirit, but at the same time let us be charitable and ever mindful of all the beauty which surrounds us, of all the grace which is in our fellowmen and perhaps in ourselves. Let us destroy the things which are evil, the ugliness which mars our dwelling places, the injustice which we do to others, above all the lies which cover all sins, but let us beware of destroying or hurting even the smallest of the many things which are good and innocent. Let us defend our traditions, all the memories of our past, which are our most valuable heritage.

To see things as they are, — by all means! But the highest aspirations of my soul, my nostalgia for things unseen, my hunger for beauty and justice, these are also realities, and precious ones. The many things which are beyond my grasp are not necessarily unreal. We must always be ready to reach out for these untangible realities which give to our life its nobility and its ultimate direction.

Ex oriente lux, ex occidente lex. Let us discipline our souls, and be loyal to objective truth, yet heedful of every phasis of reality, whether tangible or not. The scientist who is not too proud, who does not assume an aggressively "western" attitude, but remembers the eastern origin of his highest thoughts, who is not ashamed of his ideals, — will not be more efficient, but he will be more humane, a better servant of the truth, a better instrument of destiny, a gentler man.

III

THE HISTORY OF SCIENCE AND THE NEW HUMANISM

MY SECOND lecture has given you, I hope, a broader view of the history of science than is usually entertained. Most people think of it as something very technical and very arid, which may be, according to their own feelings for science, either very attractive, indifferent, or forbidding; just as a history of chess might interest a chess player considerably, but the attitude of most people toward it would be one of indifference or boredom. The history of science which would not refer to technicalities would be very incomplete indeed, but it goes far beyond these technicalities however important each one of them may be. It is the history of civilization, not only the few latest centuries of it, but the whole of it from the earliest times, as deeply as we can penetrate them, down to our own; it is the history not only of ourselves and our friends, our province, our country, our continent or our race, but of all countries, of all people, North and South, East and West. It is the history of mankind, only a part of it to be sure, but the essential part, the only one which can explain progress across the ages.

Permanent progress (if these words can be put together) can always be derived in one way or another from the attainment of knowledge or the discovery of some application of it. Human genius knows other forms of creation than the purely scientific one, but it knows no other, adequate to its steady improvement and perpetuation. What is the good of conquering if one cannot colonize? of gaining anything if one cannot hold it? Knowledge is the main conqueror and the only colonizer and administrator. Napoleon is reported to have remarked: "The only conquests which leave no regrets are our conquests over ignorance," and he knew what he was speaking of for he had made conquests of all kinds. They are also the only ones which can be continued and improved indefinitely, that is, as long as man wills it.

Humanists of the old type are fond of showing that many men of science, even among the most distinguished ones — the very ones who are called the greatest authorities on this or that — are so uncultured, so ignorant outside of their specialty, so lopsided, that their claims of leadership are preposterous. Granted, yet in most cases this is not due to any inadequacy of science, but to the fact that our education is so stupidly arranged that scientific subjects and the so-called "cultural" ones exclude one another instead of being brought into harmony. And if the old humanists were listened to, that horrible situation would not only continue but become considerably worse. As science grows in complexity and in actual size, the evils of specialization — of premature specialization without sufficient background, — must necessarily increase. Independently of any system of education, there are also narrowminded men, — and some scientists, even eminent ones, belong undoubtedly to that unfortunate group; it is as if their minds were preternaturally acute but only in the straitest region. Such scientists will continue to exist, but science cannot be blamed for their occurrence. If some of them make important discoveries and enrich us all to that extent, let us be grateful to them, and forget their lack of intellectual grace, as we would forget some kind of physical disability or ugliness. The literary people who are so quick to notice the shortcomings of scientists should bear in mind that there are of necessity mediocre men — a good many of them — in every field. There is however an important difference. The efforts of mediocre scientists may be occasionally of considerable value; indeed there is much scientific work which requires a very elaborate technical training and yet is so repetitious and tiresome that it must necessarily disgust original minds; such work must be done, and honest plodders, with little if any imagination, can perhaps do it best. Can we say the same of mediocre writers or artists? Are the authors of obscene or silly books, the painters who transform our exhibition halls into insane asylums, equally harmless? I am afraid they are not. If the mediocre scientist is a dull god, the mediocre artist is a mad one.

It is a common error to think of all scientists as if they were built upon the same pattern. Aside from their intellectual size which varies exceedingly, their types are many. One great division in this field as well as in others is between romanticists on one side, and classicists on the other. The former jump from one subject to another, change more than once and rather suddenly their direction, habits, and methods, and behave somewhat capriciously; the others spend their lives walking steadily with infinite patience and energy toward the same goal. The history of science contains perhaps less romantic figures than the history of art, but far more so than most people imagine. Think of Thomas Young, of Arago, of Galois, not to speak of the more erratic examples of earlier times. On the other hand, the external lives of many artists are as quiet and uneventful as that of the average business man. Think of Brahms!

The main opposition in the intellectual world is not, however, between romanticists and classicists — universal as this is — but rather between mystics and rationalists.[1] That is, between the men who claim to know the truth by a sudden and fundamental intuition, and those who assume on the contrary that it can only be reached by a very slow, tedious, and difficult process involving abundant means of guiding and checking every step of it. The former maintain that their knowledge is of a deeper kind than can be obtained by humbler efforts, and that they can

[1] These words "mystics" and "rationalists" are defined in the following sentences.

even reach the absolute; the latter, less ambitious, admit that though their knowledge is gradually improving in extent and precision, it remains always imperfect and relative. I would not deny all value to mystical knowledge, but unfortunately its validity can never be checked and hence it is for all practical purposes inexistent, for one cannot depend on it; in the second place, it is extremely limited and of its nature unprogressive. It reaches the absolute so fast that it seems unable to hold anything else. The latest mystical writings are not appreciably different from the earliest, which is not very surprising after all, since the absolute cannot be more so now than it ever was.

It is well that there be a few mystics among us to remind us of the relativity of our knowledge, and of its indigence with regard to the essential problems of life. Rather than leave these problems unsolved, which is intolerable to them, the mystics create solutions which satisfy them but cannot satisfy others. The rationalists prefer to be ignorant than to pretend; they prefer no knowledge to one which is of such frail and evanescent nature that it cannot be shared with other men. The mystics may be the salt of the earth — I don't mind their believing that they are — but it is the non-mystics who build up the world, not only the material world but the world of thought, and keep it in order; it is they who accomplish mankind's purpose.

While the mystics' efforts have been singularly

sterile, except perhaps insofar as they themselves were concerned (and for this we have to accept their own statements), those made by the men using scientific methods have been fertile beyond the wildest dreams. It does not follow that the latter were less disinterested than the former, but simply that their methods were better; the chances are that their disinterestedness, even as their humility, was greater. The disciplined intellect of the man of science did not simply conquer the material world; it gave us a revelation of the immaterial one far transcending that which had been given before by poets and dreamers.

The more I think of it the more convinced I am that disinterestedness is the keynote of the best scientific efforts. This disinterestedness is due mainly to the feeling of participation, — of conscious participation — in the mysterious activities of the universe. The man of science who has the real *feu sacré* in him feels that though he is but an infinitesimal part of the whole, yet his own endeavor may contribute, however little, to the fulfillment of man's purpose: a deeper understanding of nature, a closer adaptation to it, a better guidance, a more intelligent devotion. And this may be also the fulfillment of a greater purpose. If the essence of religion is an earnest consideration of life, independent of any selfish or personal motive, if it is the clear consciousness of the unity and wholeness of life and of our integration with it, then the pure scientist is intensely religious.

Similar ideas were beautifully expressed by Huxley when he said, "Science seems to me to teach in the highest and strongest manner the great truth which is embodied in the Christian conception of entire surrender to the will of God. Sit down before fact as a little child, be prepared to give up every pre-conceived notion, follow humbly wherever and to whatever abysses nature leads, or you shall learn nothing. I have only begun to learn content and peace of mind since I have resolved, at all risks, to do this." [1]

You understand, entire surrender of self: that is what I call disinterestedness; nothing less. One cannot help feeling that any such disinterested effort must increase the sum of good will in the world. When life seems a bit hard, the man of science can always preserve his equanimity by concentrating his thoughts upon the truth, far away from all the tattles and the futilities of life. In that field no deception will await him. I hasten to add that in such a case his equanimity would be tinged with sadness instead of being, as it should, a source of pure joy. To be truly happy and gay we must be able to pursue the truth, not alone, but among lovable men and women, who are kind to us and to whom we can show our own kindness. Even as the discovery of any particle of truth, whether it be to our advantage or not, pleasant or unpleasant, is a positive gain for the whole world, even so every act of kindness is a creation in the right

[1] Letter to Charles Kingsley, Sept. 23, 1860. Printed in his Life and Letters by his son, Leonard Huxley (vol. 1, p. 219, 1900).

direction. Need we prove it? Does not the daily experience of every one of us amply confirm it? There are a number of scientific facts and theories of which we are as sure as one can be of anything, though we know they are only approximate images of reality. We are not less certain that kindness, whether we be the recipients of it or the actors, enriches our life, and that unkindness or indifference impoverishes it.

It would thus seem that the natural line of conduct for a scientist would be to search for the truth, and when he has found it, to purify it and himself as much as possible, and always to be kind. Let us be kind even to those who condemn science without knowing anything about it.

Science like religion implies disinterestedness, earnestness, austerity. At its very best it leads to a kind of sanctity, as illustrated by the lives of such men as Faraday and Darwin. However it is unwise I believe to compare it with religion, and even more so to try to make a religion out of it. It is perhaps better not to speak at all of the sanctity of science. Science and religion are two domains which may occasionally overlap, yet remain separate. It is equally confusing to oppose them and to mix them. Science is neither philosophy, nor religion, nor art; it is the totality of positive knowledge, as closely knit as possible; it is as different from its practical applications on the one hand, as it is from idle theorizing and blind faith on the other. It behooves us to make no extravagant claims for it, and to be as humble as we can. Those who are too proud of their knowledge, the chances are that their knowledge is shallow or new or both. Let those who don't know brag and prophesy; those who do know prefer not to speak too much or too loud. The greatest scientists have generally assumed an attitude of humility, which was not always devoid of ostentation, for they were human and imperfect. Witness Lord Kelvin saying at the end of a life as full of discoveries as one could wish for and which to the average person must have seemed the limit of success: "One word characterises the most strenuous of the efforts for the advancement of science that I have made perseveringly during fifty-five years; that word is failure. I know no more of electric and magnetic force, or of the relation between ether, electricity, and ponderable matter, or of chemical affinity, than I knew and tried to teach to my students of natural philosophy fifty years ago in my first session as Professor." [1]

However precious the fruits of science may be, and they have proved to be infinitely precious on every plane of life — from the most utilitarian to the very highest — they are but of small value as compared with the spirit which brought them to light. That spirit is not new; it is almost as old as man himself; it informed the whole development of science from the very crudest experiments of early man down to the boldest deductions of modern physicists. The

[1] From Kelvin's address at the time of his jubilee (Glasgow, 1896), as printed in his Life by Silvanus P. Thompson (vol. 2, p. 984, 1910).

noble emperor Marcus Aurelius spoke of it when he said: "Nothing is so conducive to elevation of mind as the ability to examine methodically and honestly everything which meets us in life, and to contemplate these things always in such wise as to conceive the kind of universe they belong to, their use and their value with regard to the whole. . . ."[1] "To examine methodically and honestly" is exactly the scientist's function, but it was far more complicated and difficult than Marcus Aurelius, and even the greatest scientists and philosophers of antiquity could imagine. In fact one might claim that the history of science is the history not so much of discoveries as of the method which made them possible; for the method, which is the womb out of which issue all the discoveries past, present, and future, is naturally more important than any one of these can possibly be.

Now it is not enough to obtain knowledge, nor even to purify it as much as is in our power — and so to say to climb to the very top of every peak — we must also humanize it. As I have shown in my first lecture, this is preeminently the historian's task, for how else could we evidence the deep humanity of science, if not by explaining its concrete and lowly origins and the endless vicissitudes of its development? It is also the historian's privilege to make

[1] Οὐδὲν γὰρ οὕτως μεγαλοφροσύνης ποιητικὸν ὡς τὸ ἐλέγχειν ὁδῷ καὶ ἀληθείᾳ ἕκαστον τῶν ἐν τῷ βίῳ ὑποπιπτόντων δύνασθαι καὶ τὸ ἀεὶ οὕτως εἰς αὐτὰ ὁρᾶν ὥστε συνεπιβάλλειν, ὁποίῳ τινὶ τῷ κόσμῳ ὁποίαν τινὰ τοῦτο χρείαν παρεχόμενον τίνα μὲν ἔχει ἀξίαν ὡς πρὸς τὸ ὅλον. . . .

younger people appreciate the value of the earlier efforts, however crude they may seem, and to implant admiration and reverence into their minds. It is so easy to scoff and be superior and cynical! Make it clear to them that if they cannot admire past efforts, because the results have ceased to be admirable in the light of our latest knowledge, far from proving any superiority, they will only betray their own smallness. A man's moral worth is largely a function of his capacity for admiration and reverence.

The teaching of the history of science, however important it may be, is in itself insufficient to inculcate the points of view which I have explained and which may be grouped under the convenient heading of "The New Humanism." A few lectures or even a courseful of them would give the students some conception of the new spirit but hardly more. Thus far our education has been almost exclusively literary, the scientific courses, however numerous, remaining so to say on the outside of it. Science teachers are not expected to impart any education, but simply to teach their own technicalities. College administrators often speak of scientific and cultural courses, which is a neat way of expressing the same invidious distinction. Is it not clear that as long as science is not meant to educate, it must somehow fail to educate? To break that vicious circle a complete revolution in our system of education will be necessary.

Though this may seem to take me far away from

my own subject, it is essential at this point that I should sketch the new system I have in mind.

The basis of any education is the knowledge of one's own language. One can never know it too well, and one's whole life is hardly long enough to master it. Few are those who will continue indefinitely to study it, but at least the schools of every grade ought to give their pupils a far better send-off in this respect than they generally do. The full possession of a language (even of one of the smaller ones) is the foundation of personal culture, and this will apply just as well to scientific as to literary culture if one admits at all that distinction. It may be desirable to know more than one language, but the superficial knowledge of no matter how many will never replace the deeper knowledge of a single one.[1] The study of language is naturally correlated with all other intellectual efforts, for one cannot study anything without linguistic means, and, excepting manners and tricks, our knowledge — our objective and transferable knowledge — cannot be more precise than the language expressing it. Our very expressions give a measure of the clearness and refinement of our thoughts, and vice versa if our thoughts are clear we shall be able to express them more clearly; if they are subtle, our words will evidence their subtlety.

[1] As the reader may possibly be surprised by some outlandish expressions of mine, I must warn him in fairness to myself that I am now writing in a language which I had to acquire relatively late in life, and which is not — and in all probability will never be — my very own.

This correlation instead of being hidden and jeopardized by the stupid dualism of our education, should be emphasized in every possible way, by the science masters as well as by the others.

Pedantic scientists or other pedants may deplore the fact that our languages are such complicated organs, so full of capricious irregularities and inconsistencies that it requires far more time and energy to master them than would be the case if they had been built with better logic and greater economy. However, this weakness is also indirectly a source of strength and beauty; a natural language, however regular its structure, is not a geometrical abstraction; if it ever was, it ceases to be so in the process of being used; it gradually acquires all the graces of a living thing. More exactly it was ever a living thing, and all the complex regularities of its structure are comparable to those in the anatomy of plants and animals; they are independent of any conscious purpose; consciousness was introduced by grammarians only at a very late stage when the language was already perfected and when literary masterpieces had already been produced. The natural genesis of language increases its difficulties tremendously, but it also increases its subtleties, its mysterious charm and elegance, its possibilities of expression. So it is that the study of language is endless and the reward of it, equally endless. For example, we do not really know a language until we have become familiar with all the masterpieces it has produced and which in their turn have transformed it; nay, we do not know it until we have used it ourselves in a great variety of circumstances, to pray and love, and to express all the feelings of our souls.

Its language is the most valuable possession of a people, and it is the more precious because its blessings are fully open to each man according to his own merits. The greatest treasures of their ancestral literature are within the reach of the poorest men. Mind you, here is again verified my saying that every cultural progress is ultimately derived from scientific efforts, for cheap printing was made possible only by a whole series of discoveries based on positive knowledge. And this is illustrated once more in our own days by new developments of infinite importance. The discovery of printing and all the improvements that have been gradually added to it have brought abundant writings, the best as well as the worst, into every home, and each person, however poor, has thus been given the possibilities of reading what his soul was yearning for. Yet the written word is not the real thing; writing is simply a means of storage and transmission; the real language is the spoken one. Down to our own time, if the best written words were truly available to all who wanted them, the same could not be said of the spoken ones. Only the relatively few people whose privilege it is to move in polite circles were ever able to hear good talk well spoken. Once more science has come to our rescue and raised us up to a higher level. The discovery of radio makes it now perfectly possible to distribute the best spoken language even in the most remote places of the earth. The poorest boy could read Shakespeare if he wanted to badly enough; it is now (or it ought to be) possible for him to hear the immortal plays acted by great actors; it is as if his books had become alive and were speaking aloud to him. The fact that the radio, as the printing press, is often ill-used is no argument against the invention itself, nor against science, but only against the foolish and wicked men who turn blessings into curses. It will take a considerable time for men fully to deserve the wonderful tools which science is giving them with such abundance. Immense opportunities are created by science every day; it is for us to improve them, and we must not blame science for our failure to do so.

Time after time people have dreamed of establishing a new artificial language which would be as simple as the natural languages are complicated. In fact various so-called international languages have been created, but even if one of them did supersede the others, it would never suppress the natural languages, and hence it would simply mean the addition of a new language to all the others of which there are already far too many. Moreover if any artificial language were sufficiently popular, who could guarantee that it would not develop according to the genius of the people who would use or abuse it? After all it would be spoken by human beings, not only by fanatical grammarians. If it were sufficiently international, it might even develop along divergent ways, as did Latin and Arabic.

I do not believe in the necessity of a new artificial language, least of all for the people who have the inestimable privilege of speaking already one of the great languages of the world. But it is almost a necessity for the people speaking the smaller languages,[1] to study also one of the larger ones; and it is almost a matter of duty for all men who pursue their education beyond the primary stage to become acquainted with a second language. However large the number of languages (it is almost endless), the number of the greater languages — of those which one might call, because of their wide diffusion, the world languages — is relatively small, say, five or six. It is clear that if all the people who enjoyed a high school education knew two languages (including at least one of the greater ones), and if all the college men and women knew three (including at least two of the greater ones), it would be easy to converse with one's fellowmen almost anywhere. I may add that as a man steeps himself in an extra language he increases indirectly the understanding of his own, he becomes gradually conscious of many subtleties which he did not appreciate before because he took them too

[1] Needless to say, the adjectives small and great as applied to languages refer simply to the number of people using them. I call small a language spoken by relatively few people; great, one spoken by many people. The intrinsic value of each is out of the question.

much for granted. Moreover each natural language opens a new horizon; it helps one to know more intimately another people, another culture, and thus by comparison to be more sensitive to the qualities and to the defects of one's own inheritance.

The idea of an international language is utilitarian yet impractical; it is really anti-cultural, but its power for evil and its power for good are equally small. However little time may be needed to study such a language, that time is wasted. Moreover the mastery of any vocabulary, whether natural or artificial, will always require a long and continued effort; not simply to know the words when one sees them, but to have them at one's fingers' tips will always be the main task in the study of any language. When our vocabulary is sufficiently large and fluent, it helps us to solve grammatical difficulties almost unconsciously.

We do not need an international language of the same kind as our natural languages, for these answer their every purpose perfectly, but we do need what might be called international languages of very different kinds, and there are three of these of which every person should know at least one or two. I take now the word language in a broader sense, a means of communicating one's thoughts to other men. These three languages are mathematics, music, and drawing. They are international, or let us say, essentially human in the deepest sense. The mathematical symbols and conventions now have currency all over the civilized world; they mean the same things everywhere with a precision unequalled in any natural language. The study of mathematics is the best means of learning to think with rigor and without ambiguity. I need not explain the merits of drawing and music and how the lives of the people, who can understand these languages or better still speak them themselves, are enriched. It is as if two new worlds were opened to them. Moreover these three languages do not in the least duplicate the ordinary languages for they answer different purposes and serve to express thoughts which could not be expressed otherwise. The teaching of music, drawing, and painting ought to be far more common than it is, and more efficient. To be sure its aim would not be to create more artists, for we can only support a very few of these at a time, the very best, and there may already be too many of them as it is, but simply to open more eyes and ears to all the unspeakable beauties of the visible and audible worlds, to multiply men's chances of communication with other men and women, to increase their humanity and their happiness.

Those who might be afraid that the system of education I am advocating would be too scientific, will please notice that thus far I have not spoken of any science except mathematics. For some strange reason this last subject obtained the favor of the old humanists, as opposed to the other branches of science. From their point of view the study of mathematics was as unprofitable in a material way and hence as genteel as that of Greek, while the very mention of chemistry would have made them shudder with horror. To be sure the man who studied chemistry with no thought but of money did not educate himself into the bargain; nor did the gold digger who struck "pay dirt" become by that very fact a man of culture, even if it paid abundantly.

Some amount of scientific knowledge should be taught to boys and girls of all ages. This teaching might be begun as soon as they can read and write, and be gradually increased according to circumstances. It can be made as simple or as complicated as one may wish. The smallest children, if they have any intelligence, can be taught without pains, gaily, an astounding amount of science by a sufficiently gifted instructor. One may begin with the purely descriptive parts which require only some power of observation and memory. Thus a good deal of astronomy, geology, anatomy, botany, zoology, etc., can be explained in a very simple way. The fundamental methods of science can be illustrated by means of easy experiments, and thus will the spirit of experimental philosophy gradually inform the minds of the pupils. It is of course this that matters above all. It is perfectly possible — and it is not even difficult — to inculcate the scientific spirit, as I have defined it in these lectures, upon children in various proportions according to the age and intelligence of each. High school pupils should have the opportunity to review the main facts and theories of many branches of science. No attempt should be made to teach too many facts; there is no point in that. One fact well understood, if possible by means of personal experiments, is worth more than a hundred learned by rote. In the meanwhile, the pupils would become more familiar with the experimental method, in various forms, and their mathematical technique would be steadily improved.

Finally some historical teaching should be spread over the whole schooling period, in very small doses to begin with, the amount being gradually increased as the pupils grew. In contradistinction to most educators I believe that the proper appreciation of historical facts requires if anything more intellectual maturity than that of scientific facts. Indeed historical events are created by the conflicts between man and circumstances, they are the results of many passions and one cannot really understand them if one has not experienced these passions in one's own heart. The political history of Greece is not essentially different from that of our own times and to appreciate it to the full one must have taken a share in the political struggles through which contemporary society is passing on its way to a better order of things. The history of civilization I would put back as far as possible in the curriculum, and it would be of course focused upon the development of science. By that time the students would already be sufficiently familiar with the scientific spirit to be interested in the history of its conquests, and perhaps their knowledge of human nature would also be sufficient to make them appreciate the abundant personal implications of the history of science, its rich and deep humanity. A teacher endowed with a modicum of imagination could make them realize the pathetic greatness and all the joys and miseries of man's pilgrimage throughout the ages.

Thus far I have spoken of education such as it is imparted in our schools, but I have in mind the whole of education from the cradle to the grave: For it does not begin later nor end earlier. Nor have I spoken of the special schools whose purpose is professional rather than cultural. I cannot insist too much upon the necessity of introducing a number of historical courses in these schools. A student of law, for example, should be well acquainted with the history of it from the earliest times down to his own, otherwise his understanding of pure law will never be complete, nor his professional standards quite as high or as fine as they ought to be. A student of medicine should know the history of it, or he is doomed to remain an uncultured physician. In every case, the history of the science or the art concerned affords the best means of bridging the awful gap between technicalities, professionalism, commercialism on one hand, and disinterested culture on the other. The need of humanizing the professions is even greater than that of humanizing science.

I have not dealt with the learned languages. This is not due to any prejudice against them, neither is it due to ignorance. Most of the adversaries of science have practically no knowledge of it, except for the disjointed fragments which they may pick up in the newspapers; and as they are unable to put these fragments together and reconstruct any single body of doctrine, their knowledge remains worthless, and more likely to confuse than to enlighten them. One may question the honesty of their attacks on something foreign to them. In fact their behavior is caused by lack of intelligence rather than of honesty; it would be proper to compare them to the rustics who blame the ways of "foreigners," without bothering to know what those ways may be, save that they are foreign and different, and hence wicked. No such objection can be made to me, for in the course of my studies I am obliged to use frequently four learned languages and though I am not a specialist with regard to any of these, I know each sufficiently to enjoy my use of it.

I have not included the learned languages in the program sketched above because I have reached with the utmost reluctance the conclusion that they will have to disappear from the education of the average boys and girls. This will mean a loss, but not as severe as one might think for it cannot be denied that the number of pupils who attain a sufficient knowledge of Latin or Greek to read these languages with pleasure is very small. In the majority of cases that knowledge is nothing but bluff, like these false façades which architects build sometimes to make you believe that there is a building behind them. Yet even as it is, the lack of Latin and Greek will be a real loss. But what is one to do? It is not possible to go on adding one subject to another indefinitely. Having to choose between a little more mathematics and science and a better knowledge of modern languages, especially of one's own, on the one hand and Latin and Greek on the other, I do not hesitate to sacrifice the latter, because it will be less needed and also because in the majority of cases the knowledge ultimately attained will be inadequate for any practical use. If something must go it is the bluff. A good knowledge of any language, learned or not, will not fail to raise an individual to a higher plane; but a mock knowledge can but leave him on his old level; it will increase neither his humanism nor his usefulness, only his conceit.

To be sure the abandonment of ancient languages as a compulsory or general subject, will not put an end to their study, but it will restrict it to much fewer students and purify it. These few electing to study Latin or Greek, or both, will be presumably better qualified for it, and it will be possible to make them progress faster and to obtain better and more durable results. It ought to be made clear for example, that no language can be considered known as long as one's knowledge of it is purely passive. This remark applies of course with special force to the living languages; a reading knowledge is really but half a knowledge, for one cannot know accurately a language which one does not write, nor fluently, one which one does not speak. The real thing is the spoken language, the written one is but a pale image of it. But the remark applies also to the so-called dead languages: they ought not to be dead, and they would not be dead if the teachers were fully alive to their responsibilities. No one should be allowed to teach any language who cannot write it with relative ease and accuracy and who cannot speak it with some volubility.

Furthermore some pupils should be encouraged to devote their attention to Semitic languages — Arabic and Hebrew — rather than to Latin and Greek, and should begin their study early enough in life to permit the attainment of a moderate proficiency. Whenever possible these languages should be studied in the localities where they are still used in one form or another.

In other words instead of enticing thousands of pupils to study languages in which they have no real interest, it would be better to consider such study as a privilege which relatively few deserve and which they can only deserve by their adeptness. There would be less sham Latinists, but there would probably be more genuine ones, and these, being put on their mettle, would be far better. The change would be a fortunate one for all concerned, including the Latinists.

Thus my program could be summed as follows: Deep study of one's own language, and possibly of one or two others. Mathematics, drawing, and music. The main scientific facts and theories; the experimental methods; the scientific spirit. The history of civilization.

I have hardly spoken of the history of science, though it is understood that the history of civilization would be largely focused upon it. The fact is that the history of science, however important, is far less so than the subjects I have mentioned. I am not at all hypnotized by my own studies. I believe that the history of science should be studied very deeply, far more so than it has been heretofore, but as in the case of the learned languages, I would not try to enforce its study upon too many people. Those sufficiently attracted to it and qualified to undertake it should be encouraged, and the very fact that a group of people would dedicate themselves to it in earnestness would in the course of time influence the teaching of both science and history considerably. It is probable that a time will come when courses on the history of science will be offered in almost every college, but even if such courses were not offered, a curriculum organized along the pattern which I have outlined would already go a long way toward the fulfillment of the new education.

As opposed to the old humanists who would deliberately increase the gap between science and the humanities, the main purpose of that new education would be to bridge the gap and to close it as much as possible. The solid literary and artistic basis and the insistence on the historical point of view even in the scientific courses would oblige the more scientifically minded to consider more carefully the nonscientific aspects of life; on the other hand the frequent explanations of the scientific method by men familiar with the history of science and with all the vicissitudes of human progress, would enable the more literary minded to understand the spirit of modern civilization. To be sure such results could not be attained at once, nor even to-morrow, for they would necessitate the existence of instructors able to unite the scientific and the historic points of view, and such instructors cannot be produced on short notice. However, I can imagine a time when no one will be allowed to teach history whose scientific ignorance has disqualified him to understand its inwardness.

It is clear that the most urgent task is the organization of "normal" courses on the history of science, and of seminaries in one of our larger colleges; in fact, the creation of a whole department devoted to these new studies. This department would be the nursery and the headquarters for the study and teaching of the history of science in America. As science becomes more complex and is subdivided into more and more branches, as its study involves greater difficulties and expenditures, it stands to reason that no college, except perhaps the very biggest, can continue to provide instruction and organize research in every subject. Why should they? Of course there ought to be a number of introductory courses on the main subjects, but with regard to more special investigations the intellectual needs of the country would be served best if there was more concentration of research. One university would become the main center for organic chemistry, another for astrophysics, a third for mediæval history, a fourth for Egyptology, etc. The bigger the university the more such centers it might possess, but it would be foolish to attempt to be the center of everything. This determination of the main scientific interest of each university might conceivably be arranged by common agreement. Each college would be lifted up to a higher level by its more thorough exploration of a definite field. The prestige of each and of all would be heightened with a minimum of wasted energy.[1] According to that general plan, one of our colleges might take special trouble to organize the study and the teaching of the history of science. This could only be done by one of

[1] To prevent misunderstandings, let me add that my plan would not necessarily involve the sacrifice of any subject which is now being taught, but while the present policy of most colleges is to divide its resources as fairly as possible between the different departments, taking into account the needs of each, according to the new policy, one department would be deliberately advantaged. For example, a relatively small college might decide to pay special attention to Egyptology, and would promote special investigations in that field. While it could not hope to have a library adequate to every purpose, it would try to gather as complete an Egyptological library as possible. Etc. The main point is that the whole college would eventually benefit by its supremacy in a single subject.

the largest colleges because of the synthetic nature of these studies; the department itself would not be very expensive but its efficiency would depend upon the existence in close proximity of as large a library as possible, and of many collections, museums, and laboratories.

Let us now consider more in detail the organization of the department which I have in mind.[1] To begin with, the study and teaching of the history of science subdivides itself naturally into two groups, first, the study of the whole history in chronological order, what we might call the history of civilization focused upon the growth of knowledge, secondly, the study of the development of separate branches of knowledge throughout the ages, for example, the development of mathematics, or of a single branch of mathematics. The first group is less technical from the purely scientific point of view, but more so in other respects (archæological, linguistic).

A general course on the history of science could be organized in many different ways. I shall give my own conception of it for the sake of example. I would divide such a course into four or five independ-

[1] The following paragraphs are borrowed almost verbatim from my third article on the Teaching of the History of Science (*Isis*, 13, 272–97, 1930). As that article was written by myself and published in my own journal, I do not bother to indicate more exactly the extent of my borrowings.

ent semestrial courses of thirty to forty lectures each, as follows:

(1) Antiquity
(2) Middle Ages
(3) Sixteenth and seventeenth centuries
(4) Eighteenth century
(5) Nineteenth century, with references to the twentieth.

The two last ones might be united into one. The whole course would thus cover four or five semesters.

The four (or five) parts of the course should be independent to enable students of different kinds to follow only one or two of them if they chose. Only a student planning to make a special study of the history of science, and possibly to devote his life to it, would be expected to take the complete course, and in addition he would have to attend seminary exercises and to carry on investigations under the instructor's guidance. It would be better for him to take this course in its chronological sequence but if it so happened that he entered his junior year at the time when the instructor had reached the third or fourth part, he might begin then and study parts one and two later. This would not be as harmful as it might seem. Even so a scholar might study profitably American history first, and Greek history later, though the opposite order would be preferable.

A classical student would be advised to take the

first course, and it would be indeed very helpful to him. It is unpleasant to come across some of them — not only students but teachers — whose ideas of ancient science are extremely rudimentary. Their ignorance of the subject is sometimes so deep that they are not ashamed of it. And yet he who does not know Greek science has not really penetrated the Greek miracle. The development of Greek medicine or geometry is as admirable as that of Greek tragedy, philosophy, or sculpture. But it has this additional interest that it is somewhat easier to account for it by connecting it with earlier developments which occurred in other countries. For example in the case of mathematics and medicine, we can, if not reconstruct, at least evoke two millennia of earlier experience.

In the same way every mediævalist should be advised to take the second course. In a sense this course would be even more useful to him than the first to the classical scholar. Indeed mediævalists have given us an entirely distorted view of the Middle Ages, because of their failure to consider the evolution of positive knowledge and of technique, and to take into account the enormous intellectual activity of Islām and Israel, and the valuable contributions of India and China. Most mediævalists are Latinists, and they have led us to believe that the Latin side of the Middle Ages was, so to say, the whole show, when it was in fact relatively insignificant for many centuries. From the eighth to the eleventh century the main intellectual efforts were made under the patronage of Islām; in the twelfth century considerable efforts were still made by Muslims and Jews. After that the importance of Christendom increased by leaps and bounds, but the achievements of individual Muslims and Jews remained very important, and one cannot discard them without seeing the total growth of mankind in a wrong perspective.

Similar remarks might be offered with regard to the other courses. The third might interest students of European history, also those devoting themselves to art and literature, while the fourth and fifth would appeal especially to students of science and philosophy.

So much for the general course, but such course (or courses) should be completed by a series of special courses devoted to separate sciences. I have already indicated the essential difference between the history of science as a whole on the one hand and the histories of particular branches of science on the other. In the general courses attended by students whose scientific background is as heterogeneous as possible, it is clear that technicalities would have to be reduced to a minimum. The instructor should relate the history of only such ideas of which he could give a sufficient explanation without digressing too far; this would oblige him to eliminate certain topics, notably mathematical ones. Hence the general course would remain, at its best, somewhat incomplete and vague from the purely scientific point of view. As opposed to it, the histories of special sciences would be far more concrete, more technical, and more rigorous.

On account of that essential difference, the qualifications of instructors would be different too. To put it briefly, those giving the general course should be better historians, those teaching the special histories, better scientists. It would necessarily happen that the instructor in charge of the general course would speak of discoveries which were outside of his immediate experience. On the contrary no one should be allowed to teach the history of mathematics, who was not a trained mathematician, nor the history of medicine, who was not an experienced physician. I am not thinking so much of theoretical knowledge which an intelligent and energetic layman could possibly obtain, but of the practical knowledge which is largely esoteric, and which is in some ways more pregnant, because it is nearer to life. To appreciate the finest points of medical evolution, one must have been brought up in the atmosphere of a medical school, of hospitals, one must have acquired a physician's outlook. It stands to reason that he who would teach the history of a science, should be as familiar with that science as possible. If not, his course — even if technically correct — will be full of little misconceptions and will finally mislead students, instead of guiding and inspiring them.

It would be difficult to say which is most necessary — the general course or any of the special ones — because they answer such different needs. But the usefulness of special courses, great as it is, is restricted for each of them to a much smaller group of students.

It is, I believe, essential — for educational and even for technical reasons — that a physicist should have some knowledge of the history of physics, but such a knowledge is less necessary to other students, except those planning to become historians of science.

By the way, the necessity of such special courses was clearly realized years ago by the Belgian government, which prescribed that nobody could obtain a doctor's degree in any science without having satisfied the examiners that he had some knowledge of its history. Unfortunately this remained hardly more than a pious wish because the teaching of that history instead of being intrusted to specialists was abandoned to the mercy of scientists, few of whom were sufficiently interested to make a genuine attempt to master it. One must bear in mind that the average scientist is generally too much engrossed in his own problems to pay much attention to history, even to the history of his own discipline. Most of the Belgian courses, I am afraid, were superficial and amateurish, yet even that was better than nothing.

The principle was undoubtedly excellent. No one should be recognized a master in any subject who did not know at least the outline of its history. Of course it would be foolish to expect him to have any deep historical knowledge, but he should know the main landmarks and the leading personalities, — he should be acquainted with his scientific ancestors.

This is almost a moral obligation. We might compare it to the obligation for any educated citizen to know the history of his country. The obligation is of the same kind and of the same order of magnitude in both cases. That is, we don't expect the average American gentleman to be highly versed in American history, but we would be shocked if he betrayed ignorance of the great events. We have a right to expect every scientist to have a similar knowledge with regard to the very science to which his life is devoted, and which constitutes so to say his intellectual fatherland. For a physicist not to be sufficiently familiar with Galileo and Newton, is just as shocking as for an American not to know Washington and Lincoln.

It would not be practical, nor would it be necessary, to organize such courses for every science, but they should be provided for every main group. For example, in a large university eight courses should be offered dealing respectively with the following sciences (or groups of sciences):

(1) Mathematics
(2) Astronomy
(3) Physics
(4) Chemistry
(5) Biology (including Psychology)
(6) Geography and Geology
(7) Anthropology, ethnology, sociology
(8) Medicine.

The subdivision of these courses, and their very titles, might vary somewhat according to local circumstances or to the personalities of the teachers. For example, psychology might be dealt with in the history of biology or in that of medicine (together with anatomy and physiology), or in a separate course, or again it might be included in the history of philosophy. I suppose this last solution has often been chosen, and by the way, is it not remarkable that the history of philosophy has been taught elaborately for ages, and that the history of science must still fight for recognition?

While it would be almost a crime to intrust a general course on the history of science to a professor lacking a thorough historical training, these special courses might be committed, at least in the beginning, to scientists having a modicum of historical curiosity.

It should be noted that one or two of these special courses might be taken care of by the instructor in charge of the general course in accordance with his own scientific training. For example, I have given on various occasions, the first two courses (History of mathematics; History of physics), and if I had time, I would gladly undertake to teach the history of astronomy or chemistry, but I would not like to teach the history of biology or geology or medicine (though I deal with such subjects in my general course) because my own experience in these fields was not sufficiently direct nor sufficiently long.

A total staff of three to five professors (i.e., instructors of different grades) would be needed to give the five general courses and the eight special ones (i.e., thirteen semestrial courses), plus some more advanced ones, and to conduct the seminaries completing them.[1] This would involve a not inconsiderable expenditure, but some of it could possibly be charged against the special departments concerned. For example, the cost of a course on the history of physics would be almost insignificant as compared with the total maintenance of the department of physics. Moreover as I shall try to show presently, the organization of these historical courses would transform and purify the intellectual atmosphere of the scientific, historical, and philosophical departments, and would bring them closer together.

Let us now consider the qualifications of teachers. Scholars to whom a course on the history of science is intrusted should be primarily scientists, having an intimate knowledge of the experimental method. Generally speaking a mathematician would not be properly qualified; of course he might be admirably qualified to teach the history of mathematics, but not to teach the history of science, because his knowledge of the experimental method is likely to be inexistent, or purely theoretical like that of a philosopher. It does not matter in which special field the experimental technique has been acquired but the genuineness of that technique matters very much. For example, the carefully prepared experiments which are made by

[1] Indeed it would not be necessary to offer every course every year.

students in laboratory courses are not in themselves sufficient; it is necessary to have made at least some experiments into the unknown.

While any sound scientific preparation would be good enough to form the teacher's scientific background, a physiological one would perhaps be the best, as it would imply a preliminary training in a great variety of subjects, and some familiarity with the ever-recurring conflict between the mechanical and biological points of view.

At any rate the teacher must be primarily a scientist, for a course on the history of science would be an evil thing indeed if it became the vehicle of false or inaccurate scientific ideas. But he must be also, if secondarily, a historian and philosopher. Of course his very interest in the history of science suggests a certain amount of historical curiosity, it proves the assumption of a historical attitude — which is essential — but it is a question whether that attitude is genuine or not and well understood by himself. Such an attitude is very complex for it involves an inclination to penetrate origins and to unravel chronological sequences, a willingness to replace ideas in their own background before attempting to judge them, the faculty of seeing the past live in the present and the present in the past. Moreover he must have a sense of historical accuracy, causing him to suffer from historical errors, at least from avoidable ones, with the same intensity as from scientific or logical ones. Nowhere does the immaturity of our studies appear more clearly than here: scholars who would feel dishonored by the commission of scientific errors which they could have easily avoided, indulge in historical statements with the utmost levity.

The philosophical qualifications need not detain us long, for they are obvious enough. The historian of science must have a good epistemological training. Even if he has no intention of discussing epistemological questions, he must be aware of their presence. He must be able to discuss the logical as well as the chronological sequence of scientific facts, and to examine and compare scientific methods. Above all he must have a sufficient power of generalization to recognize the cardinal facts and the main theories and give them due prominence in his teaching. The relative importance attached to facts and theories by contemporary opinion is often wrong, and it is one of the historian's duties to correct it in the light of further events. A similar remark applies to men of science. In this field as in any other, the contemporary classification of men is often at fault. Men who were deemed very important seem much smaller, and others who were neglected and despised, much greater, when their lives and works are contemplated from a distance. The correction of such injustice is the historian's privilege, but it puts a heavy responsibility upon him. It is the main test of his historical sense, and also of his common sense, of his wisdom.

Of linguistic qualifications it is hardly necessary to speak. It is almost a truism to state the need of a reading knowledge of German, French, and Italian. This will always be necessary. Other languages may be needed according to the special investigation in hand, the most important being Greek, Latin, Arabic, Hebrew, and the European languages not yet mentioned. At present I am thinking of the teacher, who would already be well off if he knew German and French, not so much of the special investigator whose needs vary considerably and are properly endless.

One of the reasons why he must be primarily a scientist is that scientific knowledge, being essentially systematic, must be acquired in an orderly sequence. It cannot be built up bit by bit, haphazardly, but only in a methodical way, like a monument. This costs time and thought, and it requires careful guidance. It should be noted that a great part of science, notably the descriptive part of it and much of the experimental technique, can be learned to great advantage in early youth. For historical knowledge I believe the opposite to be true. First, the order of study is much more flexible. For example one can thoroughly understand the history of America without knowing that of Egypt. Second, the more mature we are, the better we are able to appreciate it. How could children understand the passions and conflicts of grown men, and yet it is these which form the very texture of history.

It is better to have some experience of life before learning its history; even so, it is better to obtain some intimate knowledge of things and men, and then we shall be naturally interested in the history of that knowledge.

I have spoken thus far only of the intellectual requirements. For a proper teaching of the history of science a number of material accessories are equally necessary. Such teaching should be given in a hall equipped for the making of simple physical, chemical, and biological demonstrations. This implies the possession of various apparatus, some of which, it is true, might be borrowed from other laboratories as the occasions for them arose. Moreover the teacher ought to have good sets of wall charts, maps, models, etc. Many of these charts and specimens would be of the kind required in the scientific courses, but some at least would be of a purely historical nature. For example to explain the discovery of the circulation of the blood, it would be well to have a chart illustrating the erroneous Galenic views, as well as the usual one. Cheap models of ancient instruments would also be exceedingly useful; e.g., the navigation instruments used by the early navigators, the apparatus of the alchemists, the first microscopes and telescopes, etc. It is clear that such tools would help not only to make matters clearer to the students, but also to fix them in their memories.

However the main advantage of these paraphernalia is that their very presence would oblige the teacher to eschew vague and worthless generalities, to stick to scientific and historical realities, and to introduce the handicraft part of science which is perhaps the most fertile. For though science is a creation of the human mind, the mind could never have developed it unaided by material devices. Libraries and museums are like extensions of our memories, and even, as far as they are classified, of other parts of our intelligence. But I was not thinking of them, rather of the instruments used by natural philosophers. Much of our experimental knowledge — and with few exceptions the best of it — developed almost naturally together with the instruments which human hands had built and used. It is truly essential to exhibit this interdependence of intellectual processes and handicraft, of minds and tools. This is the only way that one can realize the very life of science, what one might call its own reproduction power. For the minds of men are inspired and guided by the instruments which other minds have invented, and each instrument is like a center of intellectual crystallization. And so on: the younger scientists improve the instruments bequeathed to them, and better instruments make better scientists.

The main purpose of all these courses would be best attained if they were organized in a special department, which would be so to say a "liaison" department between all the others. One of the chief defects of the present system of college education is its fragmentariness. This seems an unavoidable consequence of scientific progress, and of course the fragmentation which we are now witnessing is only a beginning.

Maybe a hundred years from now pulverization will be a better word for it. At all events, the situation is bad enough as it is. A student taking a number of courses, say, English, Greek, Mathematics, Chemistry, can hardly imagine that these subjects are related. To him they are utterly disconnected; nor are his teachers more comfortable about it, for they speak different languages, think in different ways. The Greek professor knows no chemistry and is rather proud of it. The mathematician has forgotten what little Greek he ever knew and is not aware of any loss. The chemist has long since reached the conclusion that historians and philosophers are windbags and nothing more. How could they understand one another? And if one of their students has an especial longing for intellectual unity and suffers from this disintegration and this kind of drifting, how can they, who hardly appreciate what their colleagues do, allay his anxieties?

And yet all the subjects which are taught in our colleges are deeply united, — branches of the same tree, twigs of the same branches, leaves of the same twigs. To realize it, it almost suffices to think of the past, to replace oneself in an earlier stage when knowledge was less differentiated. The twigs may not know one another but they belong to the same tree, as you can easily see if you go down from twig to branch, and from smaller branches to bigger ones. The Greek miracle was named Hippocrates or Archimedes, as well as Æschylus or Phidias. It is really the Greek spirit, the spirit of Aristotle and also to a smaller extent that of Plato, that we are fighting for. A chemist, who understood this thoroughly, would be more of a humanist than the Greek professor, if the latter did not (which is but too often the case).

The department of the history of science would keep in touch with all the other departments; it would be its very mission to do so. Parts 1, 2, 3 of the general course would put it in communication with the classical and historical departments; parts 4 and 5, with the scientific departments. However the main bridges to the latter would be the special courses. For example, the one on the history of mathematics would be organized with the intimate collaboration of the mathematical department. It would even be an essential feature of the latter, for whatever his special interest, every mathematical student would be advised to take the historical course in his junior or senior year.

The grouping together of all these scientifico-historical courses would be preferable for other reasons: it would facilitate their coordination, their proper distribution between a few specialists, and would make the team work of the latter more effective and more pleasant.

Above all, the purpose of that department being to reestablish synthesis and unity, it is clear that this would be best accomplished if it were not identified with any one of the departments to be reunited. For example, if these courses were organized by the philosophical department — as might well be done — there would be a double danger: first, that the teaching would be too philosophical, would become in fact something different; second, that the confidence of the historical and chiefly of the scientific departments would not be obtained as easily. Scientists have not forgotten that it took them at least a thousand years to fight the hegemony of theologians and philosophers, and that their own immensely successful career did not really begin until that hegemony was broken. On the other hand if these courses were offered by the historical department it might happen that the philosophic or even the scientific interests — which are predominant — be neglected, or that instructors be appointed whose scientific qualifications would be insufficient, and the scientists would drift away from them instead of coming nearer.

There is still this to be considered. Sooner or later it will be necessary to make possible the obtention of master's or doctor's degrees in this new field, the history of science. If such degrees could be granted only by either the historical or the philosophical department, it is probable that the requirements would be too severe with regard to the historical and philosophical knowledge, and too indulgent with regard to the scientific knowledge — which is truly essential — and to the history of science proper.

The main requirements of such degrees would be:

(1) Sound knowledge of one branch of science (including experimental work) and of its history.

(2) Lesser knowledge of two other branches and of their history; one of them being sufficiently distant from the main scientific subject. For example, if the main subject were physics, the two others might be astronomy and general biology.
(3) Linguistic abilities to be determined in accordance with the candidate's purpose. A reading knowledge of French and German would be required in every case.
(4) Knowledge of historical method and of general history.
(5) Deeper knowledge of a certain historical period, or of the history of a certain race or people.
(6) Epistemology and logic.

It may be observed that the history of science has this in common with other historical branches, that while it requires highly trained specialists to promote our knowledge of it, every educated man can enjoy its fruits. In other words, the courses on the history of science are difficult to give, but easy to receive. To give them truly well, no amount of knowledge would suffice, wisdom and intuition would be equally necessary. But if they were truly well given, not only would they appeal to all intelligent students, but they would help them to understand better the rest of their curriculum, and to make a wiser use of it. It would at once humanize and integrate their knowledge.

There is one curious misunderstanding which I have not yet touched, that is, the confusion between the ideas of "history" and "introduction." This is due to the fact that one of the best ways of introducing a scientific subject is to explain its birth and early development. This is so true that elementary textbooks on, say, astronomy, have sometimes been called "history (or the story) of astronomy." This confusion is readily understandable if one realizes the many ambiguities of our word history.

However helpful the historical point of view may be to introduce a scientific subject — to make people understand, for example, what astronomy is and what astronomers do — it does not help us very long. If one were to study a branch of science in the historical order, one would never get to the end of it, and, what is worse, one's knowledge would be badly knit together, and of a very precarious nature. The almost infinite complexity of science would soon bewilder the strongest minds, if it were not steadily counteracted by the synthetic efforts of some scientists. The result of these efforts is to reclassify the items of knowledge in an order which is generally very different from the chronological order, or from the one thought to be the most natural; indeed it may seem extremely artificial. It is often found that the best way of teaching a subject to adult minds is to begin by explaining some of the latest and most abstract notions and using these as the "elements." Thus the final order of the subject taught is almost the opposite of the chronological order of invention. For example, a professor of chemistry would (or might) find it convenient to explain at the very beginning — as if these were the most obvious matters — our highly complicated views on the structure of atoms and spectra and the principles of thermodynamics.

Under the influence of the confusion between "history" and "introduction," when courses on the history of science have been improvised in some of our colleges, they have generally been offered to the youngest students, the idea being that such a course would be for them a general initiation and would help them to determine their curriculum.

It is possible that such an "initiation" course is useful — this does not concern us at present — but it is certain that it is very different from the course on the history of science which we have in mind. As explained above, one cannot be genuinely interested in the history of something which one does not know. On the contrary, the more extensive their scientific knowledge, the more would the students appreciate the historical survey which would be offered to them in their junior, or preferably, in their senior year. Far from being introductory, this course should be as far removed as possible toward the end of the curriculum. After having studied many things, each as it were in its own separate box, and in a systematic, logical order, they would review their intellectual acquisitions from the historical point of view. This would gradually bring out the natural but deeper relationships obtaining between objects apparently disconnected; it would help students to weave all their knowledge into a single fabric, and incidentally to fix it in their memory. The course on the history of science would not be in any sense introductory, but rather a sort of conclusion, of final coordination. Call it an initiation, if you please — as every general course is — it would not be a scientific but rather the supreme humanistic initiation.

After having explained the organization of the study and of the teaching of the history of science, which forms so to say the core of the new movement, we may now return to the consideration of the movement itself in its broader aspect. To put it briefly, the New Humanists wish to associate themselves with every creative activity, to help mankind go forward with enthusiasm, but at the same time to look backward with gratitude and reverence. The forward going and the backward looking are equally essential; in fact, they are complementary. It is the combination of both which is perhaps the main characteristic of the New Humanism; the combination of youthful energy and curiosity with reverence for the past. It implies a continuous struggle on two opposite fronts, against iconoclastic technicians and crude materialists on one side and against blind and futile idealists, the chickenhearted humanists of the old school on the other. We must go forward without fear, yet preserve all the sacred traditions which are our most precious possessions, our very titles of nobility; we must explore the mysterious domains which surround us and climb higher and higher and yet transmit to posterity all that was best in the past. The New Humanism is a double renaissance: a scientific renaissance for men of letters, and a literary one for men of science.

The time is ripe. Literary people, artists, philosophers, excepting a few irreconcilable old fogies, realize that science has come to stay and that its growth, however exuberant, is but a beginning. They do not wish to emulate the good Mrs. Partington who tried to keep the Atlantic Ocean out of her house with a mop. Is it not wiser to accommodate oneself to ineluctable circumstances? Moreover they realize that science is considerably more than mere technique, that however important its applications, its intrinsic value is greater still, and finally that one cannot dispose of it by saying that it does not concern man. Every scientific idea, however esoteric, is thoroughly human from its birth to its perfection. To deny its inherent humanity because in its final shape it is represented by a lifeless abstraction, would be as stupid as to deny the humanity of a poem because we know it only in cold print. Science is as instinct with life as any other product of man's activity; and as the special activity which gives birth to it is one of the very highest, it is instinct with the highest and purest life. In fact it is largely because of its extreme purity that many people, infirm as they are, fail to recognize in it any kinship with their own miserable little dreams. I repeat it is no longer necessary to explain all this to most literary people; they are fully aware of it, and ready to go half way to meet gentle scientists and try to understand their moving spirit and their ideals.

On the other hand, the more intelligent and educated scientists have lost the assurance and conceit which did them so much harm some fifty years ago. From the end of the last century on, one revolutionary discovery after another has shattered their complacency and obliged them to revise almost every one of their theories. The positivistic philosophy has become untenable; it is no more possible for our scientists to predict the frontiers of knowledge than for the old humanists to determine the boundaries of humanity. The conception that reality must necessarily be simple has been destroyed, then reestablished time after time. The appearances are certainly very complex. The whole of our knowedge, in spite of its being far more precise in many respects, and far more extensive, has become less dogmatic, and has gained a fluidity and a gentleness which were utterly lacking in earlier days. This is due at once to its own immense progress and to the influence of religious and artistic ideas. Thus the best scientists are also willing to go half way and meet their literary brethren, if these are like-minded.

It is true that the same circumstances have driven some scientists back into metaphysics, and caused a few of them to distill extraordinary visions far transcending their experience. I think the historical or humanistic attitude is wiser. There is no point in attempting to describe what may be seen from the top of the mountain when we are only half way up. Is it not better to enjoy all the beauties which are unveiled to our eyes, and to continue our ascension in a joyous and humble spirit? We do not set presumptive bounds to our knowledge; some limitations may be inherent in our weakness yet even these are gradually perfectible. What is already knowable we must try to know as well as possible, and suffer our failures, whether they be temporary or not, with patience. It is equally foolish to ignore things which we could know and to pretend that we understand others which are yet beyond our ken.

Human endeavors to obtain knowledge more complete and a purer truth and to destroy errors and prejudices will never cease. Whenever some available knowledge would increase justice, remedy misery or illness, diffuse beauty, — ignorance of it, far from being a virtue, is a sin or a crime. But even if it could be perfect, our knowledge would still be insufficient. We need beauty, love, and grace as much as truth.

It is man's sacred duty to strive for more knowledge, yet to keep his mind and heart open to all the mysteries that surround him. Our mental reach must always exceed our grasp, or else our growth would soon be stopped. No one ever went very high who did not stretch himself upward and strain his soul to the limit. Some ignorant fools would make us believe that knowledge destroys idealism. On the contrary, the clearer our sight, the deeper our vision. Blindness will never help us. We need lifelong ideals, as much as we need our daily bread, but these are always a function of our knowledge, as it were, irradiations of it. The more knowledge we have, the better and sounder may be our ideals; that is, if we are worthy.

The New Humanism will not exclude science; it will include it, and so to say it will be built around it. Science is our mental armature; it is also the armature of our civilization. It is the source of our intellectual strength and health, but not the only source. However essential, it is utterly insufficient. We cannot live on truth alone. That is why we say that the New Humanism is built around science; science is the core of it, but only the core. The New Humanism will not exclude science but on the contrary exploit it to the utmost; it will minimize the danger of scientific knowledge abandoned to its own technicalities; it will extol the human implications of science, and reintegrate it into life; it will bring together into a single communion scientists, philosophers, artists and saints. It will confirm the oneness of mankind, not only in its achievements but in its aspirations. The evils of the so-called "machine age" have been caused by the aloofness of the old humanists as well as by the narrowmindedness of some scientists, but above all by the insatiable greed of men of prey. This "machine age" must go, and be replaced at last by the "scientific age"; we must prepare a new culture, the first to be deliberately based upon science, upon humanized science, — the New Humanism.

IV

THE HISTORY OF SCIENCE AND THE PROBLEMS OF TO-DAY

THE best way of introducing the subject which I propose to discuss before you is to tell you two parables which I have been revolving in my mind so often that they have become as it were a part of its substance.

A general-in-chief sends one of his regiments to attack the enemy far away from his own point of concentration; he knows that the regiment will be destroyed but that it will enable him to attain his main objective. It all works out as the general had planned. The forlorn hope is cut to pieces, and thanks to that the main body of the army defeats the enemy and puts it to rout. The question is : were the men of the sacrificed regiment defeated or victorious? If you consider that regiment by itself, as a separate unit, they were utterly defeated. If you consider them as a part of the main army, they must share in its victory. I believe that the second point of view is the true one. Not only were those men members of the victorious army, but it was their immolation which snatched victory from the jaws of death. Not only were they victorious, they were the angels and the heroes of victory.

That is my first parable, and I invite you now to consider with me two men who are fighting one another. They are of equal strength and courage but the one is fair and gentle, while the other is rough, cruel and unfair. Thus the strength of the first is jeopardized by the many limitations exacted by his own conscience, while for the other there are no limitations whatsoever — anything goes. Who has the best chance of victory? The second of course. *Yet if you take mankind as a whole, it is the first who has won.*

When one reads such a book as Gibbon's Decline and Fall, one cannot help shuddering half of the time, and wondering how on earth did the "good people," the forward-looking ones, the "dreamers" ever triumph over the unprincipled barbarians, over the ruffians, gangsters and murderers — who in every level of society, from top to bottom, were weakening the Empire, and undermining like termites the whole fabric of civilization. How could the monks who were then almost the only reliable guardians of western culture, and who were trying to prepare the good life in their monasteries, how could such meek and inefficient individuals overcome all the forces of evil and darkness? And yet they did. The Roman Empire disappeared to be sure but its virtue and justification, its principles of piety, order and law triumphed over chaos. The triumph was far from complete; nothing is ever complete and perfect in this world, but the traditions of justice and goodness which survived were sufficient to preserve culture and hand it on to the following generations and finally to us. In the long run, the good man, though critically handicapped by his own moderation, had triumphed over the bad one.

Before trying to explain that paradox I must forestall an objection. My two parables speak only of war and struggle; is that truly appropriate to my message? It certainly is. Let us realize once for all that there is no peace in life, but only war; peace can only come to us when we are dead. The difference between peace-loving people and others is not a difference between peace and war, but between the objects of war. We mean to fight for justice and beauty and truth, because we know that those things which alone can give value to our life must be won, and won again and again, and that we can reach them only as much as we deserve. Virtue can be maintained only by uninterrupted effort. Peace, whether of the individual soul or of the people, can be realized only by constant struggle against its internal and external enemies. It is not something static and tangible, but only a dynamic equilibrium. Just as soon as people relax — just as soon as they take peace for granted — they slip backwards, and great may be their fall.

The history of science is the story of a protracted struggle, which will never end, against the inertia of superstition and ignorance, against the liars and hypocrites, against the deceivers and the self-deceived, against all the forces of darkness and nonsense. The history of art is the story of a protracted struggle, which will never end, against the inertia of ugliness, against all the people who prize the upholstery of life more than its harmony, and are ever ready to destroy the beauty of nature or besmirch their own nests. The history of society or government is the story of a protracted struggle, which will never end, against all the forms of tyranny, whether individual or social, against arbitrariness in human dealings, and against the exploitation of the weak and the poor by the strong and the rich. The history of mankind at its best is that of a Pilgrim's Progress, which can never be free from struggles, for the latter never end except with life itself.

To return to our subject, how is it that Christianity has triumphed (however incompletely) over paganism, morality (however imperfect) over immorality, relative justice over injustice, order over chaos? How did the saints succeed in disarming the ruffians, the scientists in overcoming the liars? How could gentle people survive long enough to transmit their gentleness and increase it — however slowly — in the face of a brutal world? Those events seem so miraculous that one might be tempted to abandon the attempt of explaining them, and speak only of Providence. Yet I believe that an explanation can be given if one does not insist on an absolutely complete one. Every human explanation, as you know, must fall short of perfection because of our own hopeless imperfection. But if we take for granted the existence in man — in some men at least — of an unquenchable thirst for beauty, justice and truth, then the ultimate victory of the defenceless good men over the bandits armed cap-a-pie and ready for anything can be explained. The miracle occurs because there is continuity of effort among the former, while the latter are always and necessarily at cross-purposes. When the aim of man is simply to be powerful or wealthy, his achievements are likely to be undone just as soon as he dies (if not earlier) because many other men have the same selfish motives and no one can obtain full satisfaction except at the expense of others. If his triumph is sufficient, he can hardly expect his children to continue his struggle, because they may desire to enjoy "in peace" the advantages which he has won and paid for so dearly. What is more he may himself begin the undoing of his own power, because he may have had time in the lull of his battles to realize the emptiness of such a victory as his. On the other hand if one's aim is to create beauty, the more beauty there is around him — created by others — the better it is, and when he dies, what he has created is added to the capital of beauty of the whole world. It is the same with justice and truth. There is indeed considerable emulation, a real struggle, in the creation of it, but such emulation is neither exclusive nor destructive. If a man succeeds in increasing the amount of justice (or decreasing the amount of injustice), he does not do it for himself alone, and others may continue his effort where he left off. However nowhere is the emulation a more complete collaboration than in the case of science. The competition is as intense among men of science as among any other men, but nevertheless they are all pulling together — not against one another — and whenever a parcel of truth is found by any scientist, he finds it not for himself, not even for his own people, or the people of a single nation or faith, but for the whole world.

The gentle but continuous efforts of good men are like the proverbial drops of water falling unobtrusively but uninterruptedly upon the same place and cutting mountains. They must necessarily defeat the erratic efforts of selfish men, however strong the latter and however weak themselves. The higher aspirations of men are sufficiently pertinacious and convergent to accomplish miracles, and it is only because of its continuity of purpose that humanity has at all succeeded in achieving a modicum of civilization.

That truth had been realized by one of the greatest conquerors in the material world, and one who had been given time to drink the cup of disillusion to the very dregs. Said Napoleon to his henchman Fontanes (could that pusillanimous creature understand him?) "Do you know what I wonder at more than anything in the world? At the impotence of force to organize anything. There are only two powers in the world, the sword and the intellect, and in the long run the sword is always beaten by the intellect."

The great Emperor had met the same paradox as we have but he had failed to explain it, and in his disenchantment he had perhaps reached an excessive conclusion. For if force is useless to create anything and cannot compete with reason, yet it may serve the function in human affairs of "giving moral ideas time to take root";[1] that is, it may truly help to organize order in the face of chaos, provided that it is not depraved by internal poisons and nullified by a selfish and stupid antagonism to the main drift of mankind, that it works with and not against the instinctive sociability of the race. With force at his disposal and without that superior instinct man would be nothing, as Aristotle remarked more than twenty-two centuries ago, but the most dangerous of the beasts. The very fact that he has never succeeded in playing that beastly part for very long, does it not prove the existence of an innate and persistent virtue in him, which sooner or later must destroy any form of immoral power?

Some historians were quick in realizing, even in days far remote from our own, that their main task should be to set forth not so much the military and dynastic vicissitudes and the pathology of mankind, as the obscure travail which prepared its gradual evolution and made possible the accomplishment of its higher purpose. At first the religious aspect of that purpose was emphasized, which was already an immense step forward. Thus the great Palestinian scholar, Eusebios, declares at the beginning of the fifth book of his church history: "Other writers of historical works have confined themselves to the

[1] That is Admiral Mahan's felicitous expression as quoted by Lionel Curtis: Civitas Dei (279, London, 1934).

written tradition of victories in wars, of triumphs over enemies, of the exploits of generals and the valour of soldiers, men stained with blood and with countless murders for the sake of children and country and other possessions; but it is wars most peaceful, waged for the very peace of the soul, and men who therein have been valiant for truth rather than for country, and for piety rather than for their dear ones, that our record of those who order their lives according to God will inscribe on everlasting monuments: it is the struggles of the athletes of piety and their valour which braved so much, trophies won from demons, and victories against unseen adversaries, and the crowns at the end of all, that it will proclaim for everlasting remembrance." [1]

How magnificent such a declaration in the mouth of a writer of the first half of the fourth century! Eusebios' exaltation was due to the fact that he had fully appreciated Constantine's recognition of Christianity (A.D. 312), thanks to which, to use his own words again, "two roots of blessing, the Roman empire and the doctrine of Christian piety, sprang up together for the benefit of men." [2] His understanding of history was prophetic though we would interpret it differently to-day. We would agree with him that the main purpose of historiography must be to em-

[1] Eusebios: The Ecclesiastical History, with an English translation by Kirsopp Lake (Loeb Library, vol. 1, 404–7, 1926).

[2] In the oration in praise of Constantine the Great which Eusebios pronounced on the thirtieth anniversary of the emperor's reign, i.e., in 336; chapter 16, 4.

phasize the main purpose of mankind, and that if one wishes to indicate the chief lines of human development it almost suffices to insist upon the creative and lasting achievements and to evoke the great men who were responsible for them. We could not agree completely on the achievements nor on the men, but well enough on the principle.

Behold all the beauties and glories of nature, but what could be more interesting among them than men themselves, and among men who could interest us more deeply than those who accomplish our destiny and justify our existence?

Behold the grandeur of art, but beyond the works of art we look for the men who produced them, and are these not more impressive than their own creations?

Behold the serene greatness of science, the wonderful succession of discoveries which have deepened our insight into all the mysteries of the universe, even beyond the reach of the most fantastic dreams. These are marvels indeed in comparison with which all the marvels of the Arabian Nights seem cheap and commonplace; but the greatest marvel of all, is it not the fact that those things were discovered and invented by men, substantially like ourselves?

Think of how puny we are, and contemptible in many ways, and yet that some of us — beings of our own flesh and blood — have added so much beauty to the universe and enabled us to penetrate its own beauties far more deeply and vibrate with it in far closer unison than was ever deemed possible. These are great achievements indeed, absolute conquests. Said Napoleon (it is strange that I should quote him twice, but his career — that demoniac welter of creation and destruction — is very instructive indeed): "The only conquests which can leave no regrets are our conquests over ignorance." We might add — "and over injustice and ugliness." We have no better reasons for pride in our manhood than those victories — pure victories — without dishonorable failures, without atrocities or lies, without anything to be ashamed of.

Eusebios was thinking primarily of saints, but if he had lived with us and shared our experience, maybe he would have welcomed at least some of the artists and some of the scientists in the communion of the saints, for all of them helped us in diverse ways in doing our supreme task and vindicating ourselves (in theological language this would be called "saving our souls"). Their creations — of them all — call for that "everlasting remembrance," which Eusebios had in mind and which is or should be the stuff of history. For in the long run the saints will defeat the sinners; they may die in the struggle but will share in the victory (remember my first parable).

In the long run, generous ideas will survive ungenerous ones, and justice, injustice. In the long run, beautiful things will outlast ugly ones. In the long run, truth will eradicate error. In the long run — a very long run mind you! . . . It is a long, long way to Tipperary, but it is a much longer one to Corinth or to Parnassus. We must not expect peace and plenty in our own life-time. What does it matter? Are not we, if we choose, members of the victorious army, an army unsullied by evil deeds?

Most of us will die long before the victory, long before any victory, big or small, but we will share in it, and indeed we do share in it already insofar as we have understood our relationship to the other men and to the universe, and done our best within our own little field. Nobody can do more than that, and nobody should be satisfied with less.

How does the continuity of scientific efforts manifest itself? Every chapter in the history of science is an illustration of it. It is true great discoveries are discontinuities, but when one analyzes them one realizes that these discontinuities are more apparent than real. The function of great men is essentially synthetic: they put together elements borrowed from everywhere and complete the building prepared by many others. This is not a disparagement: without them the building would not exist, but even they could not have built it if most of the materials had not been handy. Anyone taking the trouble, or rather giving himself the pleasure, of studying the history of science from its beginnings, may witness the infinitely slow accumulation of primitive inventions, followed with the discoveries of the Egyptians, the Sumerians, the Babylonians, the Minoans, the

Assyrians, the Persians. After a suspense caused by the vicissitudes of wars and revolutions the Greeks built upon these foundations their own amazing system of knowledge. As our archæological information increases, the Greek endeavor becomes less of a discontinuity, because we recognize in it more and more oriental elements. The stones of their wonderful constructions were borrowed from abroad; Greek science is not less wonderful for all that, its structure is as noble and impressive as that of the Parthenon, but we now realize that though it was built within a few centuries, it took millennia of continuous efforts to prepare it. And the same is true of the whole past. Each man adds his stone to the building, and sometimes an old building is broken to pieces and the old stones are used again sooner or later for a new one. Each man continues the work of his fellows — seen or unseen, known or unknown, friends or enemies; each people continues the task of the people who preceded them, and so on. The continuity is hardly if ever broken, because the materials of science are not sufficiently tangible to be capable of destruction, and because the cooperation of all men in this their supreme duty is spontaneous. They cooperate not because they want to, but rather because it is their function and destiny to do so. Neither race nor faith nor political boundaries can be an obstacle to a collaboration which involves the whole of mankind. To be sure all peoples are not equally gifted, their spirits may rise and flag, and the common undertaking may be interrupted here or there. Historians who follow the efforts of only a single people, race, or faith, and perhaps only in a single field (say chemistry), may have an impression of discontinuity, but if one looks at it from a higher point of view and a more catholic, the realization comes that there is no real discontinuity; the general task from time to time is differently distributed among the peoples of the earth; the work continues, but in different countries; it is as if mankind were working in shifts. Consider the development of mathematics; not to speak of our anonymous ancestors of prehistoric days, there were first an Egyptian shift and a Sumerian (non-Semitic) one, then a Semitic Babylonian one, then perhaps a Hindu one — that is, at least three or four oriental shifts; then a Greek one which may be called western though some of its work was done in Asia, then an Hellenistic team, more than half oriental, a Jewish one, an Arabic one, almost completely oriental, then a whole series of western shifts, Italian, English, German, and so on. Please note the alternance not only of peoples, but of races — East vs. West and West vs. East — and of faiths — Jewish, Christian, Moslem. Even so we have but told a small part of the story, for while the teams which I have mentioned were relieving one another, other teams — Hindu, Chinese, Japanese — were accomplishing similar tasks in different ways. How is it that all those efforts converge as they do? I have discussed elsewhere that stupendous fact. Common traditions might explain a part of it, but even when such community of tradition cannot be assumed, the independent efforts remain consistent to a degree, and the isolated discoveries may be still fitted in a logical sequence. This can only be explained by postulating the internal unity of mankind and the internal unity of science.

The main point is that the international collaboration of men in scientific creation is automatic and to a large extent independent of political circumstances. To be sure, social as well as physical cataclysms may wreck the work here or there, or there may appear some petty despots (deemed great by their partisans) who try to break the intellectual unity of mankind and to ostracize this or that portion of it. They might as well try to ostracize the North Wind or the Indian Ocean! They may succeed for a short time in their own section, but they are absolutely unable to interrupt even for a moment the scientific cooperation with the ostracized groups, and their futile and mean resistance is soon defeated and undone.

Science develops very much as if it had a life of its own. Great social events cast their shadows before and after upon science as well as upon other human activities; and however alive and independent science may ever be, it never develops in a political vacuum. Yet each scientific question suggests irresistibly new questions connected with it by no bounds but the bounds of logic. Each new discovery exerts as it were a pressure in a new direction, and causes the growth of a new branch of science, or at least of a new twig. The whole fabric of science seems thus to be growing like a tree: in both cases the dependence upon the environment is obvious enough, yet the main cause of growth — the growth pressure, the urge to grow — is inside the tree, not outside. Thus science is as it were independent of particular people, though it may be affected at sundry times by each of them. The tree of science symbolizes the genius and the glory of mankind as a whole.

The continuity of science appears in still other ways. Science grows steadily, irresistibly, and in general slowly though with occasional jumps or flights, the "syntheses" to which I have referred before. Its progress if not rapid is less precarious than any other; its diffusion is very slow indeed, but gentle. There is no need of propaganda in the usual sense, only explanations, reiterated explanations.

The law of equality of action and reaction is as valid in the spiritual as in the material world. Semitic exclusiveness was the root of anti-Semitism. Clerical despotism has always been the main source of anticlericalism. Every immoderate propaganda is bound to turn against itself like a boomerang. Conversions to any creed, whether religious or political, are worthless if not spontaneous, and when they are obtained by physical coercion they are worse than worthless. The diffusion of science, however, is of the most peaceful kind, and unless it be artificially entangled with irrelevant issues, it causes no adverse reaction. Intolerance and injustice can be alleviated by merciful deeds, but they can only be extirpated by the slow penetration of the scientific spirit.

The history of science is the story of an endless struggle against superstition and error; it is not a vivacious and spectacular struggle, but rather an obscure one — obscure, tenacious and slow. The resistance of science against every form of unreason or irrationality is so firm and yet so quiet, that it is almost as gentle as non-resistance would be, yet unshakable.

All of which indicates that science is the best instrument, if not the only one, to defeat barbarism and to establish on a solid foundation whatever kind of culture we have already managed to inherit, to gain or to build ourselves. Science aims at perfection, a definite kind of perfection within its own sphere, and it is thus led to grow continuously in a definite direction. Science aims at permanence, and hence in spite of its readiness to sacrifice always the imperfect to the less imperfect, in spite of its iconoclastic and revolutionary tendencies, it is the best guaranty of ultimate stability. Science is of its very essence international and inter-racial, it is thus the strongest bond of union between the peoples of the earth. It aims at unanimity, not concerning any preconceived idea, but concerning the very system which is being developed by the unconscious and continuous collaboration of all peoples in a task independent of themselves and infinitely superior to any one of their desires. Scientific work is one of the highest forms of altruism.

These words may sound very strange to the disillusioned people of our time, many of whom blame science for their troubles, and the more so because their expectations had been more foolish. The technical progress of last century had been so stupendous that they imagined that its continuance at the same speed or faster still would soon introduce a golden age. Well, that was a great mistake, for technicians can improve instruments only — they cannot change human nature. The foolish hopes of yesterday and the present disillusionment simply prove that these people have not understood the function of science.

To begin with, in spite of all its virtues science alone cannot give meaning to our life. Science by itself is not culture, though it is an essential part of it. This is obvious enough when one considers the scientific activities of our own age, that is, if we consider all of them, as we should, not only the best, but the mediocre and the perverted activities as well — the whole gamut. Science without wisdom is a very poor thing indeed, and technique without wisdom is poorer still. I am not thinking now of the men — alas but too numerous! — who are doing scientific work without a real vocation for it (the man of science without vocation is as pitiful a creature as the minister or the priest without an inner call). These wretched individuals help to darken the picture, but we must be tolerant, not only because a man, mediocre and spiritless, is not necessarily evil, but also because some good scientific work is done every day by men of that very type. This fact witnesses against science in general. There is far more room for mediocrity of every kind in the house of science than in the house of arts and letters. It does not follow that the former is meaner, but simply that many tasks are accomplished in it which require no imagination, and little virtue except honesty and fidelity.

The increasing complexity and difficulty of various techniques afford considerable scope for a technical virtuosity which can be just as admirable or contemptible as musical virtuosity. It is admirable when it is properly subordinated to ideas; it becomes contemptible when it is too self-centered and complacent. The mastery of an intricate and laborious technique is but too often a screen for intellectual mediocrity, even as rites and creeds may become cloaks for religious inanity.

The crisis which we are passing through is beneficial to the extent that it obliges us to revise and purify our thoughts on many subjects — not simply social and economic ones — but scientific ones as well. It discourages capriciousness and frivolity — which should include technical frivolity — and invites us to more concentration. American science like American life suffers from a certain amount of overtension and jerkiness. We need more concentration on single problems.[1] We need less hustling and more thinking — longer and quieter meditations. I cannot help dreaming often of that Valley of Humiliation of which Mercy says [2] "I love to be in such places where there is no rattling with coaches, nor rumbling with wheels; methinks, here one may, without much molestation, be thinking of what he is, whence he came, what he has done, and to what the King has called him; here one may think and break at heart, and melt in one's spirit, until one's eyes become like 'the fish pools of Heshbon.'" That would undoubtedly be easier, but as we cannot change our environment, we must make the best of it as it is. If we will it strongly enough and are sufficiently absorbed in our own pursuits, we can forget the noise and bustle of the world, and almost cancel them as far as we are concerned. All the unnecessary hustle can be stopped at once by any individual who knows well enough what is good for his soul. There is nobody but himself to prevent him from creating the spiritual rest and that opportunity of leisure which is the cradle of wisdom. Even the poorer man could do that if he refused to prostitute his mind to the radio or the movies and devoted the leisure time, which he would thus save, to working or walking quietly in his garden, or reading slowly a good book in his chamber. Many

[1] It gives me special pleasure to proclaim that in this respect the Carnegie Institution has taken from the beginning a position of leadership, and though the investigations which it has conducted date only from this century, their influence is already considerable, and their educational value can hardly be overestimated.

[2] Pilgrim's Progress.

people — rich and poor alike — complain that they have no time, but the time which they allow to be dissipated each Sunday by the gigantic newspapers would suffice to give them a solid intellectual pabulum if they used it more wisely.[1] In fact the opportunities of leisure are steadily increasing, but few people are aware of them (as opportunities), or able to improve them. One of the greatest social problems of our age is to teach men, who have just escaped the slavery of relentless labor, how to use their newly acquired freedom. I have no patience with "intelligent" people who claim they have no time to think; is it not rather that they have no brains or no will?

One could not repeat too strongly that no life can be intellectually sound if there be no space in it for quiet study and meditation. Of course each man needs some amusement but he should avoid the dissipation which chokes the mind without nourishing or

[1] Excellent materials are contained in the newspapers and the magazines, but their abundance, incoherence, hodgepodginess are deadly. Each item drives the preceding one off the mind. Ask readers, who have spent a whole morning reading them, what they have remembered. Precious little.

Meditation and concentration are especially helpful at the beginning and the end of the day, but many people begin and end their days with newspaperish confusion, and they give their brain no chance whatever for a clear vision.

It is often said that people have the government they deserve. That is no longer true to-day, if it ever was, for a reckless group may succeed in capturing the means of power and coerce the majority. It is true however that, in countries where the freedom of the press obtains, people have the newspapers they deserve. Indeed they have the privilege of voting for or against each paper each day, by buying some and refusing others. The newspapers of a free nation are truly representative of it.

relaxing it, he should resist any form of intellectual degradation, he should create and defend his own retreat and his own leisure.

The development of all kinds of techniques has been so rapid and so drastic that mankind has had no time of adjusting itself to it. The result is the chaos which we are witnessing to-day and which is upsetting our spiritual as well as our material world. To consider the latter first, the growth of industrial, commercial and financial methods and machines has been so reckless and merciless that large communities have been ruined by the very activities which should have made them prosperous and happy. The excessive mechanization of life has poisoned the wells of individual, familial and social happiness. The greatest problem which the statesmen of our time have to solve is the humanization of industry and labor, but it will not be easy to undo the evils of a century of technical ruthlessness and unrestrained greed, and in any case the problem is exceedingly difficult, for it is not enough to find a theoretical solution; one must be able to overcome prejudices and vested interests, to untwist wrong purposes and unmask false ideals. Moreover economic questions have become largely international, and social ills cannot be cured completely except on an international basis.

As to the spiritual chaos, it is so deep that it cannot be remedied by any single method, but this much is certain, no cure will have any efficacy which does not include the humanization of science. One must find means of integrating science with the rest of our culture, instead of allowing it to develop as an instrument foreign to it. Science must be humanized, which means among other things that it must not be permitted to go on a rampage. It must be an integral part of our culture and must remain a part of it subservient to the rest. The best if not the only way of humanizing it is to consider it historically — just as we have always considered the other cultural elements; one must study its genesis and evolution, and make people realize that the scientific achievements of each age were always, first and last, human achievements. Indeed, irrespective of their technical difficulties (which are interesting enough but secondary) these achievements were among the purest and the most glorious of their time. As long as science is looked at only from the technical and utilitarian angle, there is hardly any cultural value in it. For example, I cannot help smiling when I hear young enthusiasts exult because our universe is constantly increasing. Now it tickles our imagination to hear that the depth of the universe extends to so many millions of light-years, but there is nothing especially cultural or "uplifting" in that. The quality of our souls is independent of the size of the universe; we do not find ourselves better or worse, happier or more unfortunate because the universe is shown to be so much larger than we had fancied. However just as soon as we are given to understand the human inwardness of these facts, how they were discovered and how they affect the thoughts of astronomers, our minds are stirred up. Then we feel that we are offered something which concerns ourselves, not only the distant stars; something indeed which concerns us as intimately as a tragedy of Shakespeare, a painting of Rembrandt or a concerto of Brahms. If one considers science — not only as it is now (which after all is only its latest, not by any means its final stage) but in its becoming, if one surveys its birth and growth, its ramifications and confluences, and analyzes its vicissitudes — human vicissitudes most of them — and its struggles, defeats and triumphs, that is, if one reads the history of science, one reads the history of mankind at its best. The new realization of the richness of our past, which is thus opened to us; this new evaluation of the continuity of human efforts, and of our heritage of science and wisdom — that is humanism, though a new kind of it, including science instead of excluding it — scientific humanism if you please, or better, humanism pure and simple, humanism and culture.

It is noteworthy that these two fundamental problems, the humanization of industry and labor on the one hand, and the humanization of science on the other, are correlative. They have been entailed by the same causes, and there are so many interrelations between them, that the solution of the one cannot fail to help in solving the other. In particular no scientific humanist could feel happy in a world, however cultured it might seem to be, wherein the majority of men would be miserable and hopelessly subject to the political or economic tyranny of a few. The scientist needs protection and a certain amount of isolation from the noisy crowd, in order that he be able to do his best, but that does not mean that he wishes that crowd to be other than happy and contented. His own existence is precariously dependent on their good will, and he should be quite as anxious to educate and elevate them, as they to give him a chance. Again, human relationships cannot be cordial if there be a feeling of injustice and oppression on the one side or the other, and the love of man is the very kernel of humanism; if that kernel be absent, the rest cannot amount to much.

That reminds me of a saying of the Nestorian monk, Simon de Taibûtheh, who flourished somewhere in Syria or in Irâq about the end of the seventh century. Speaking of the greatest of all commandments, "Love God," he remarked, "is concerned with theoretical knowledge, and 'Love thy neighbor,' with practical knowledge." [1] The cynically minded might add that the love of one's neighbor is very tangible in its results while the love of God may camouflage anything you please. When a man speaks of his love of God, we cannot be certain of what he is talking. It may be something sublime, or only an alibi. Even so a man may speak of generosity because he is generous, or because he is particularly mean. Such ugly

[1] *Isis*, 24, 132, 1935.

doubts, I am sure, were not in Mâr Simon's mind. What he thought was that the love of one's neighbor is the practical foundation of religion, even as we say that it is the practical foundation of humanism. For all the grace and beauty of which the humanist is always dreaming, where would they be and how could they be if love were absent?

And yet, however essential the love of man, it is only a foundation, and the deep humanity of science is only a part of its justification and of our own. The chief aim of scientific research is not to help mankind in the ordinary sense, but to make the contemplation of truth more easy and more complete. This implies a deep conversion of the spirit which can only be achieved by a long and rigorous discipline. One must forsake all kinds of wishful thinking, and all thinking which is not constantly subject to verification and correction. One must educate oneself to become more and more experimental and objective. One must learn to conceive the truth one is aiming at and living for, as an ideal which may remain forever out of reach, but which he may and should approach more and more closely.

When this scientific objectivity is carried high enough it leads to a peculiar kind of disinterestedness which is far more fundamental than the disinterestedness of the most generous man. It is not so much a matter of generosity as of forgetfulness and abandonment of self. Every scientist (as every artist or saint) who is sufficiently absorbed in his task reaches sooner or later that stage of ecstasy (unfortunately impermanent), when the thought of self is entirely vanished, and he can think of naught else but the work at hand, his own vision of beauty or truth, the ideal world which he is creating. In comparison with such heavenly ecstasy, all other rewards — such as money and honors — become strangely futile and incongruous. Looking at it from that point of view, science is the best school of objectivity and disinterestedness, and the devoted men working in laboratories are very close indeed (though they hardly realize it themselves) to the monks and nuns mortifying their flesh in the cloisters. One can truly speak of the sanctity of science, as well as of its humanity, but it is better not to speak too much of it, for it is a subject far too confidential and too precious for expression. Also it is better not to encourage the constitution of a new class of hypocrites. If there be sanctity it will flourish best in secrecy; nobody should ever know of it, except perhaps much later.

This you will notice is a plea for an uncompromising idealism of which our age is more deeply in need than of anything else. We need to be taught a new life of the spirit, humble, gentle and free, without moroseness but without boisterousness. Scientific humanism, the new humanism, can provide the elements of it, or at least some of them.

We must investigate our traditions, not excepting those magnificent ones which have handed on to us the knowledge and wisdom of antiquity, of the middle ages and of all the centuries previous to our own. It is thanks to those traditions that we know what we know, and that we are what we are. We must familiarize ourselves with the great men to whom we owe them. There is nothing of which we can be more proud than of those traditions which constitute the essence of our culture, of our very hearts and souls, but we must not be too proud, lest we become unworthy of them. Moreover our past is a long record not of good deeds only, but also of evil ones: so many crimes have been committed (and are still being committed to-day) even in the name of high ideals. Hence our pride should always be tempered with proper shame and humility.

If we wish to survive we must anchor ourselves to some great purpose. For example we may try to continue some of our scientific traditions, or else if we be historically minded we may try to record them accurately. In any case the study of those traditions would fill our minds with respect and gratitude and inspire them with the supreme human loyalty — not to family, race, country, language or religion — but to truth. Loyalty to truth above everything else. Only a few of us can increase the patrimony of science and art, but we can all help in preserving it and honoring those who built it up.

If we succeed in taking a comprehensive view of the universe, including the human elements which are the most precious part of it, and if we cultivate a sense of respect and gratitude, we are establishing in ourselves the best conditions for equanimity. To be sure many things are bad enough in the present to jeopardize anybody's peace of mind, but we should compare the ills of the day not only with our unrealized ideals but also with the stark realities of the past; we should look backward for a reassurance of man's steady purpose and continuous, if very slow, progress, and forward for the guidance of our own trials and adventures.

The study of history, and especially of the history of science, may thus be regarded, not only as a source of wisdom and humanism, but also as a regulator for our consciences: it helps us not to be complacent, arrogant, too sanguine of success, and yet remain grateful and hopeful, and never to cease working quietly for the accomplishment of our own task.

HISTORY maketh a young man to be old, without either wrinkles or gray hairs; priviledging him with the experience of age, without either the infirmities or inconveniences thereof. Yea, it not onely maketh things past, present, but inableth one to make a rationall conjecture of things to come. For this world affordeth no new accidents, but in the same sense wherein we call it *a new Moon*, which is the old one in another shape, and yet no other than what hath been formerly. Old actions return again, furbished over with some new and different circumstances.

From the Epistle dedicatory, 6 Mar., 1638, of Thos. Fuller's "The History of the Holy Warre," 1639, but copied from the second edition, 1640.

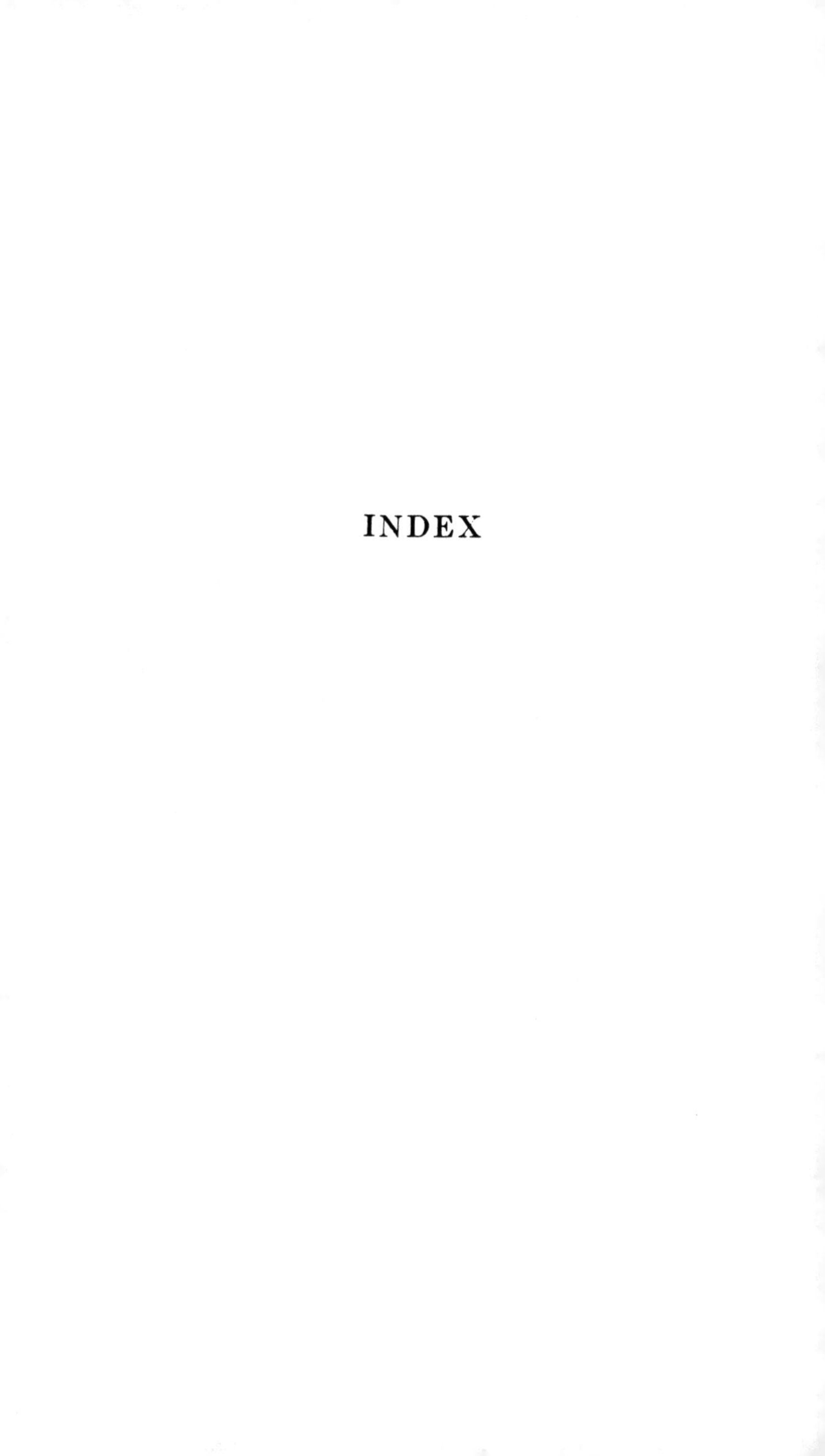

INDEX

INDEX